Systems Architecture of Smart Home Security Cloud Applications and Services IoT System

-- General Architectural Theory at Work --

William S. Chao

2

Structure-Behavior Coalescence

$$\text{Systems Architecture} = \text{Systems Structure} + \text{Systems Behavior}$$

4

CONTENTS

PREFACE

A system is complex that it comprises multiple views such as strategy/version n, strategy/version n+1, concept, analysis, designs, implementation, structure, behavior and input/output data views. Accordingly, a system is defined as a set of interacting components forming an integrated whole of that system's multiple views.

Since structure and behavior views are the two most prominent ones among multiple views, integrating the structure and behavior views is a method for integrating multiple views of a system. In other words, structure-behavior coalescence (SBC) is a single model (model singularity) approach which results in the integration of multiple views. Therefore, it is concluded that the SBC architecture is so proper to model the multiple views of a system.

In this book, we use the SBC architecture description language (SBC-ADL) to describe and represent the systems architecture of Smart Home Security Cloud Applications and Services IoT System (SHSCASIS). An architecture description language is a special kind of system model used in defining the architecture of a system. SBC-ADL uses six fundamental diagrams to formally grasp the essence of a system and its details at the same time. These diagrams are: a) architecture hierarchy diagram, b) framework diagram, c) component operation diagram, d) component connection diagram, e) structure-behavior coalescence diagram and f) interaction flow diagram.

Systems architecture is on the rise. By this book's introduction and elaboration of the systems architecture of SHSCASIS, all readers may understand clearly how the SBC-ADL helps architects effectively perform architecting, in order to productively construct the fruitful systems architecture.

ABOUT THE AUTHOR

Dr. William S. Chao is the CEO & founder of SBC Architecture International®. SBC (Structure-Behavior Coalescence) architecture is a systems architecture which demands the integration of systems structure and systems behavior of a system. SBC architecture applies to hardware architecture, software architecture, enterprise architecture, knowledge architecture and thinking architecture. The core theme of SBC architecture is: "Architecture = Structure + Behavior."

William S. Chao received his bachelor degree (1976) in telecommunication engineering and master degree (1981) in information engineering, both from the National Chiao-Tung University, Taiwan. From 1976 till 1983, he worked as an engineer at Chung-Hwa Telecommunication Company, Taiwan.

William S. Chao received his master degree (1985) in information science and Ph.D. degree (1988) in information science, both from the University of Alabama at Birmingham, USA. From 1988 till 1991, he worked as a computer scientist at GE Research and Development Center, Schenectady, New York, USA.

Dr. William S. Chao has been teaching at National Sun Yat-Sen University, Taiwan since 1992 and now serves as the president of Association of Enterprise Architects, Taiwan Chapter. His research covers: systems architecture, hardware architecture, software architecture, enterprise architecture, knowledge architecture and thinking architecture.

PART I: BASIC IDEAS

12

Chapter 1: Introduction to Smart Home Security Cloud Applications and Services IoT System

A smart home security cloud applications and services IoT system (SHSCASIS) is a system designed to detect intrusion and unauthorized entry into a house. Smart home security system [O'Dri15] consists of one or more sensors to detect intruders, and an alerting device to indicate the intrusion.

Sensors are devices which detect intrusions. Sensors may be placed at the perimeter of the protected residential area, within it, or both. Sensors can detect intruders by a variety of methods, such as monitoring doors and windows for opening, or by monitoring unoccupied interiors for motions, sound, vibration, or other disturbances. Alerting devices indicate an alarm condition. In addition to the system itself, SHSCASIS operates autonomously within networks and is often coupled with a monitoring service. In the event of an alarm, the premises control unit contacts a central monitoring station as shown in Figure 1-1. Operators at the station see the signal and take appropriate action, such as contacting property owners, notifying police, or dispatching private security forces. Such signals may be transmitted via dedicated alarm circuits, telephone lines, or internet.

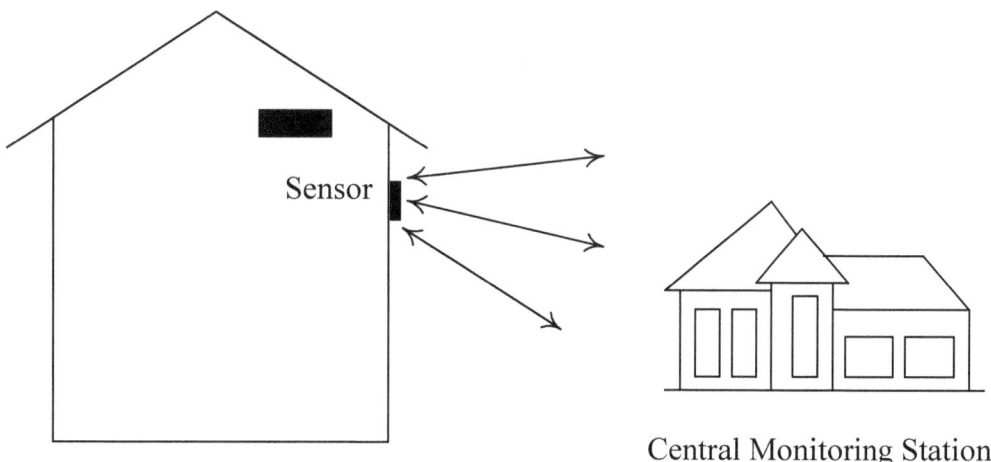

Figure 1-1   SHSCASIS Operates Autonomously and within Networks

14

A cloud applications and services Internet of Things (IoT) system is the network of physical things or objects embedded with electronics, software, sensors and connectivity to enable it to achieve greater value and service by exchanging data with the users, operators and other connected devices. Each thing or object is uniquely identifiable through its embedded cloud computing [Bern09] system but is able to interoperate within the existing internet infrastructure, as shown in Figure 1-2.

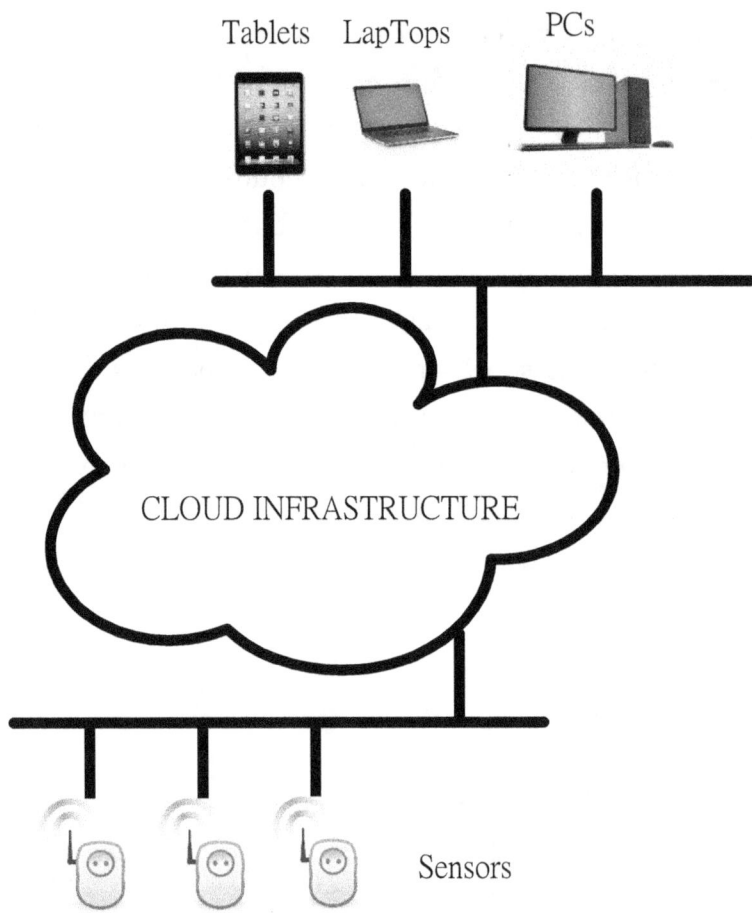

Figure 1-2   A Cloud Applications and Services IoT System

A generic cloud applications and services IoT system can also be generally represented by the following framework as shown in Figure 1-3.

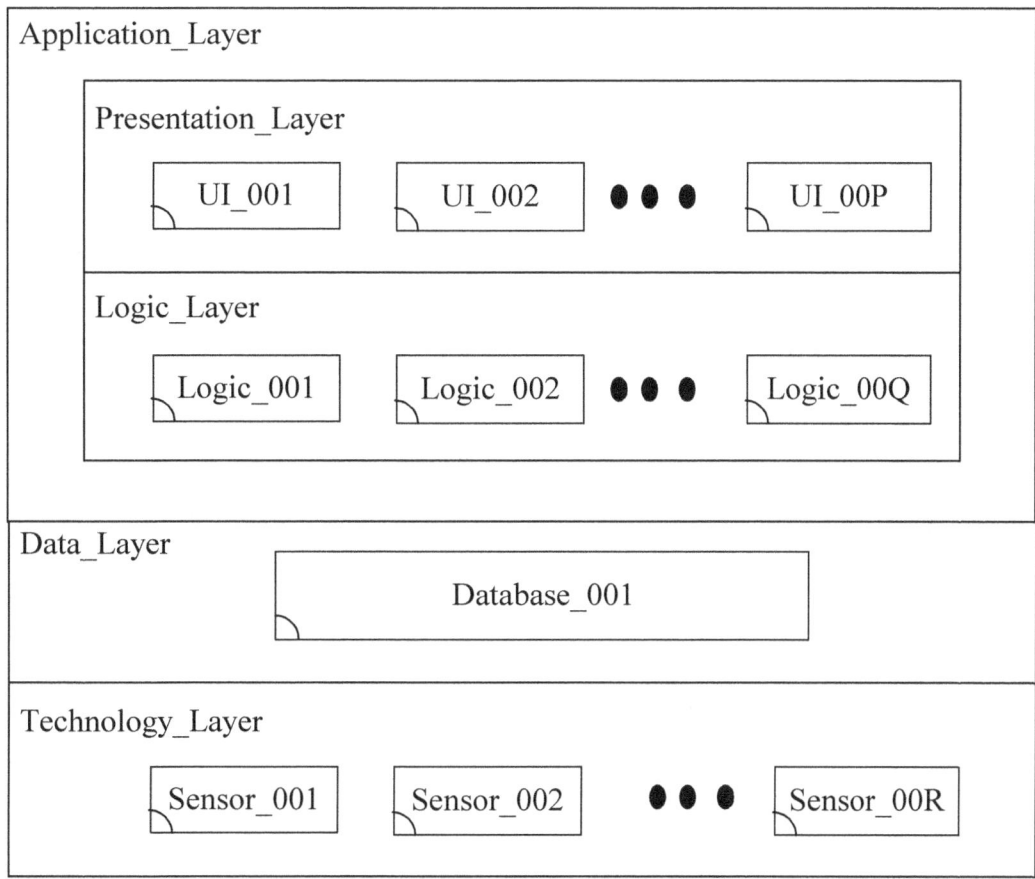

Figure 1-3   Generic Cloud Applications and Services IoT System

In the above figure, *Presentation_Layer* and *Logic_Layer* are sub-layers of *Application_Layer*. *Presentation_Layer* which describes the visual appearance of user interfaces contains many user interface (UI) components; *Logic_Layer* which describes the business logic or domain logic that encodes the real-world business rules contains many business logic components; *Data_Layer* which describes the system's logical and physical data assets contains a database component; *Technology_Layer* which describes the hardware and network infrastructure needed to support the deployment of data and applications contains many sensor components.

Advancements in cloud applications and services IoT systems present enormous potential for accurate monitoring and furnishing service to the property owners and home security providers. Behaviors of SHSCASIS consist of: a) behavior of *Registering_Home_Account*, b) behavior of *Sensing_Intrusion_Signs*, c) behavior of *Alerts_Notifying*, d) behavior of *Recording_Emergency_Response_Starting_Time* and e) behavior of *Recording_Emergency_Response_End_Time*.

1-1 Behavior of Registering_Home_Account

Each house needs to register to SHSCASIS to get the smart home security applications and services, as shown in Figure 1-4.

Figure 1-4   Registering a Home Account to *SHSCASIS*

In the behavior of *Registering_Home_Account*, a home security provider shall use the *Home_Account_Registering_UI* component to input the corresponding data for this home account registration. After that, the home account registration data will be saved to the *SHSCASIS_Database* component.

1-2 Behavior of Sensing_Intrusion_Signs

Smart home security system will be sensing the intrusion signs of each house, as shown in Figure 1-5, at regular intervals of 1 to 2 minutes.

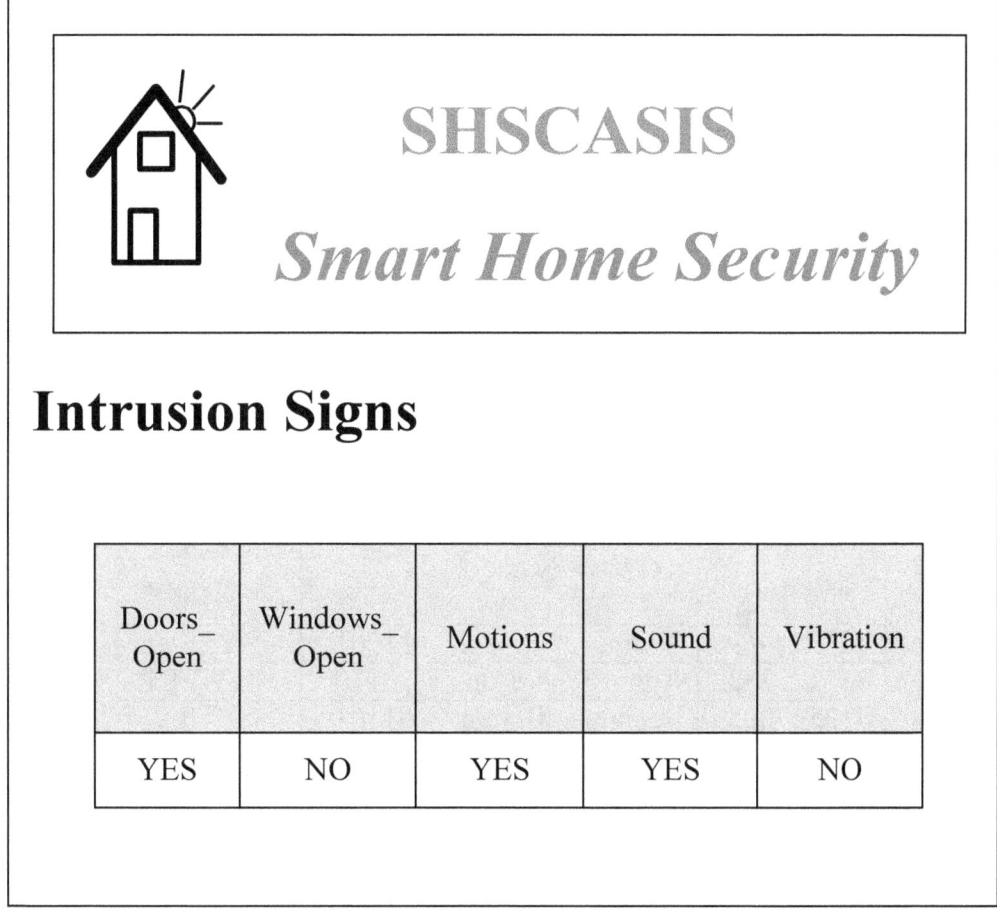

Figure 1-5   Intrusion Signs of Each House

Once the intrusion signs have been sensed, the sensor will pass the intrusion signs information to the *Intrusion_Signs_Daemon* component. After that, the *Intrusion_Signs_Daemon* component will communicate with the *SHSCASIS_Database* component to record the intrusion signs information.

1-3 Behavior of Alerts_Notifying

By constantly (about 2 minutes) analyzing the intrusion signs of all houses, an alert will be triggered if there is any abnormal situation.

In the behavior of *Alerts_Notifying*, an alerts report is displayed on the *Alerts_Notifying_UI* component screen as shown in Figure 1-6.

SHSCASIS

Smart Home Security

Alerts Report

2016/07/03, 15:30 PM

Home_ Number	Owner_Name	Alert_ Code	Emergency_ Response
A207144699	Grace Nixon	11001	YES
B411277322	Tom Richardson	10110	NOT YET
B125670055	Benjaman Bryant	01101	YES
C243112288	Lee Shelton	01111	NOT YET

Figure 1-6   Displaying the Alerts

1-4 Behavior of Recording_Emergency_Response_Starting_Time

If there is any abnormal situation, alerts are displayed on the screen for healthcare providers to take an emergency response. Emergency response is the organizing, coordinating and directing of available resources in order to respond to the alert and bring the emergency under control.

Once the emergency response has been started, a home security provider should record it on SHSCASIS as shown in Figure 1-7.

Figure 1-7   Recording the Emergency Responses Starting Time

In the behavior of *Recording_Emergency_Response_Starting_Time*, a home security provider shall use the *Emergency_Response_Starting_Time_UI* component to

20

input the corresponding data for this emergency response starting time. After that, the emergency response starting time data will be saved to the *SHSCASIS_Database* component.

1-5 Behavior of Recording_Emergency_Response_End_Time

If there is any abnormal situation, alerts are displayed on the screen for home security providers to take an emergency response. Emergency response is the organizing, coordinating and directing of available resources in order to respond to the alert and bring the emergency under control.

Once the emergency response has been accomplished, a home security provider should record it on SHSCASIS as shown in Figure 1-8.

Figure 1-8   Recording the Emergency Responses End Time

In the behavior of *Recording_Emergency_Response_End_Time*, a home security provider shall use the *Emergency_Response_End_Time_UI* component to input the corresponding data for this emergency response end time. After that, the emergency response end time data will be saved to the *SHSCASIS_Database* component.

Chapter 2: Introduction to General Architectural Theory

A system comprises multiple views such as strategy/version n, strategy/version n+1, concept, analysis, designs, implementation, structure, behavior and input/output data views. A systems model is required to describe and represent all these multiple views.

The systems model describes and represents the system multiple views possibly using two different approaches. The first one is the non-architectural approach and the second one is the architectural approach. The non-architectural approach respectively picks a model for each view. The architectural approach, instead of picking many different models, will use only one single multiple views coalescence (MVC) model.

In general, MVC architecture is synonymous with the systems architecture. Since structure and behavior views are the two most prominent ones among multiple views, integrating the structure and behavior views becomes a superb approach for integrating multiple views of a system. In other words, structure-behavior coalescence (SBC) leads to the coalescence of multiple views. Therefore, we conclude that SBC architecture is also synonymous with the systems architecture.

2-1 Multiple Views of a System

In general, a system is extremely complex that it consists of several evolution&motivation views such as strategy/version n and strategy/version n+1 views; it also consists of various multi-level (hierarchical) views such as concept, analysis, designs and implementation views; it also consists of many systemic views such as structure, behavior and input/output data views [Kend10, Pres09, Somm06].

Figure 2-1 shows that in a system all these strategy/version n, strategy/version n+1, concept, analysis, designs, implementation, structure, behavior and input/output data views represent the multiple views of a system.

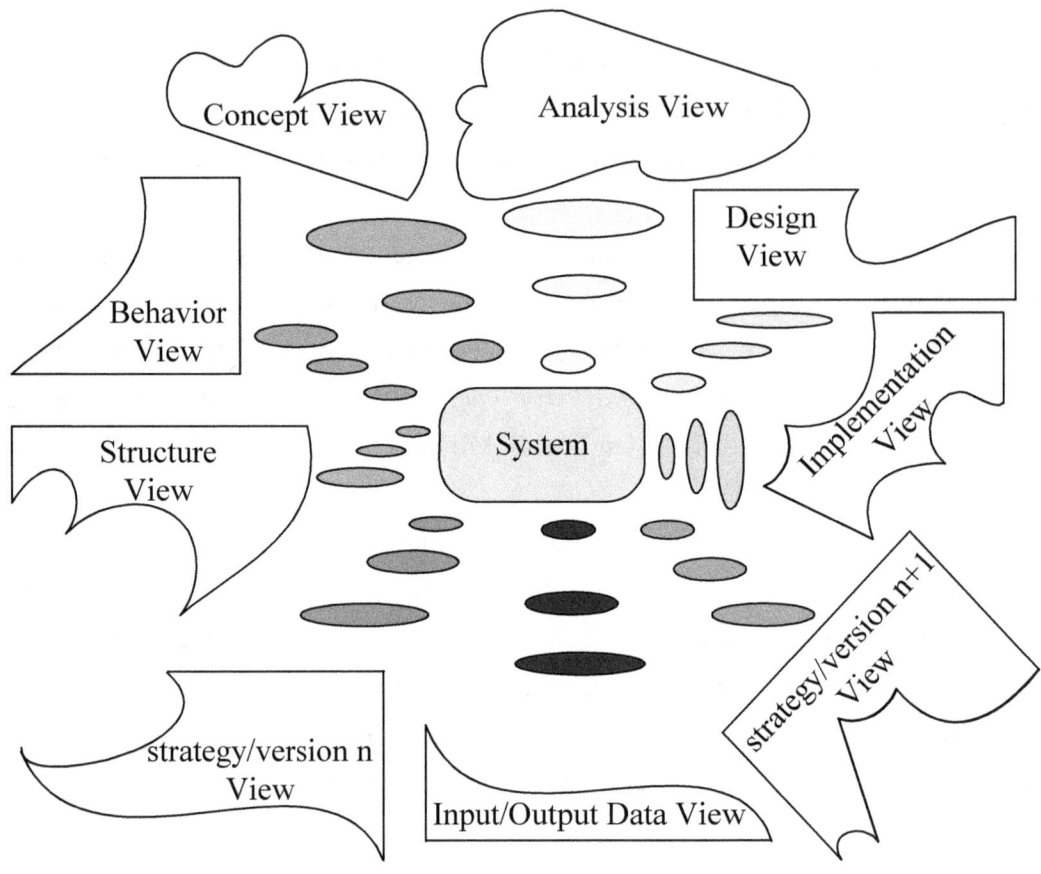

Figure 2-1   Multiple Views of a System

Among the above multiple views, the structure and behavior views are perceived as the two prominent ones. The structure view focuses on the systems structure which is described by components and their composition while the behavior view concentrates on the systems behavior which involves interactions (or handshakes) among the external environment's actors and components. Strategy/version n, strategy/version n+1, concept, analysis, designs, implementation and input/output data views are considered to be other views as shown in Figure 2-2.

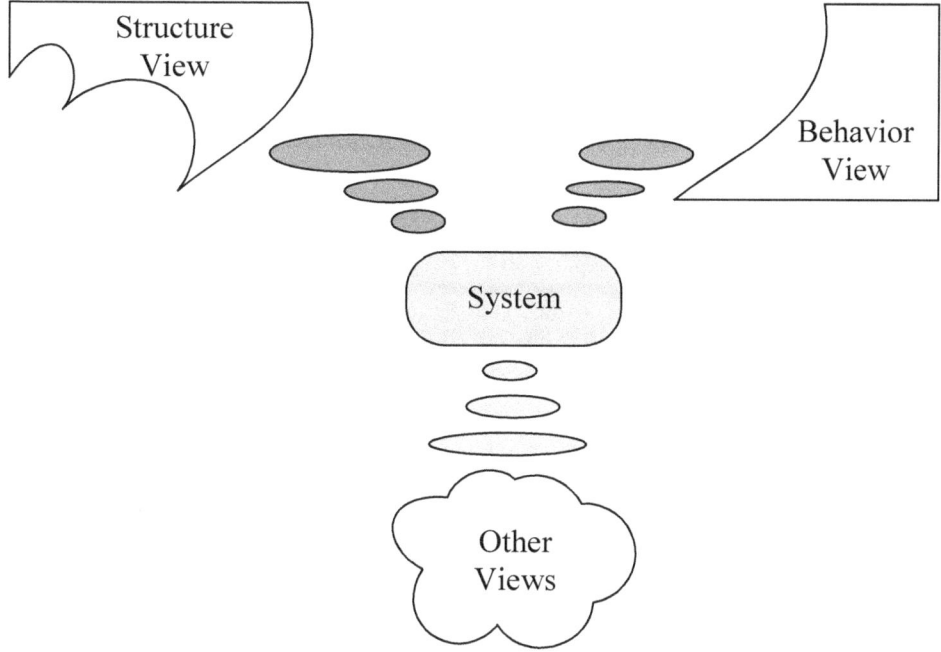

Figure 2-2   Structure, Behavior and Other Views

Accordingly, a system is defined in Figure 2-3 as an integrated whole of that system's multiple views, i.e., structure, behavior and other views, embodied in its assembled components, their interactions (or handshakes) with each other and the environment. Components are sometimes named as non-aggregated systems, parts, entities, objects and building blocks [Chao14a, Chao14b, Chao14c, Chec99].

A system, hopefully is an integrated whole of that system's multiple views, i.e., structure, behavior, and other views, embodied in its assembled components, their interrelationships with each other and the environment.

Figure 2-3   Definition of a System

Since multiple views are embodied in a system's assembled components which belong to the systems structure, they shall not exist alone. Multiple views must be loaded on the systems structure just like a cargo is loaded on a ship as shown in Figure 2-4. There will be no multiple views if there is no systems structure. Stand-alone multiple views are not meaningful.

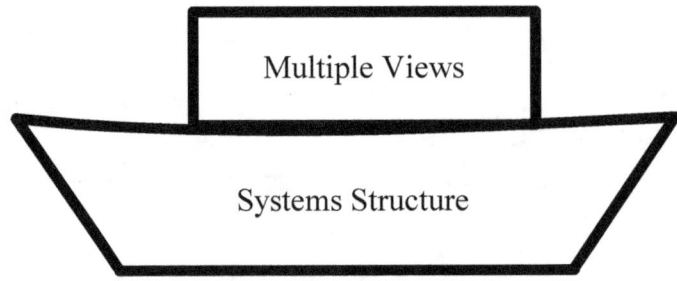

Figure 2-4   Multiple Views Must be Loaded on the Systems Structure

2-2 Non-Architectural Approach versus Architectural Approach

A system is exceptionally complex that it includes multiple views such as strategy/version n, strategy/version n+1, concept, analysis, designs, implementation, structure, behavior and input/output data views.

The systems model describes and represents the system multiple views possibly using two different approaches. The first one is the non-architectural approach and the second one is the architectural approach.

The non-architectural approach, also known as the model multiplicity approach [Dori95, Dori02, Dori16, Pele02, Sode03], respectively picks a model for each view as shown in Figure 2-5, the strategy/version n view has the strategy/version n model, the strategy/version n+1 view has the strategy/version n+1 model, the concept view has the concept model, the analysis view has the analysis model, the design view has the design model, the implementation view has the implementation model, the structure view has the structure model, the behavior view has the behavior model, and the input/output data view has the input/output data model. These multiple models are heterogeneous and not related to each other, thus there is no way to put them into a conformity model.

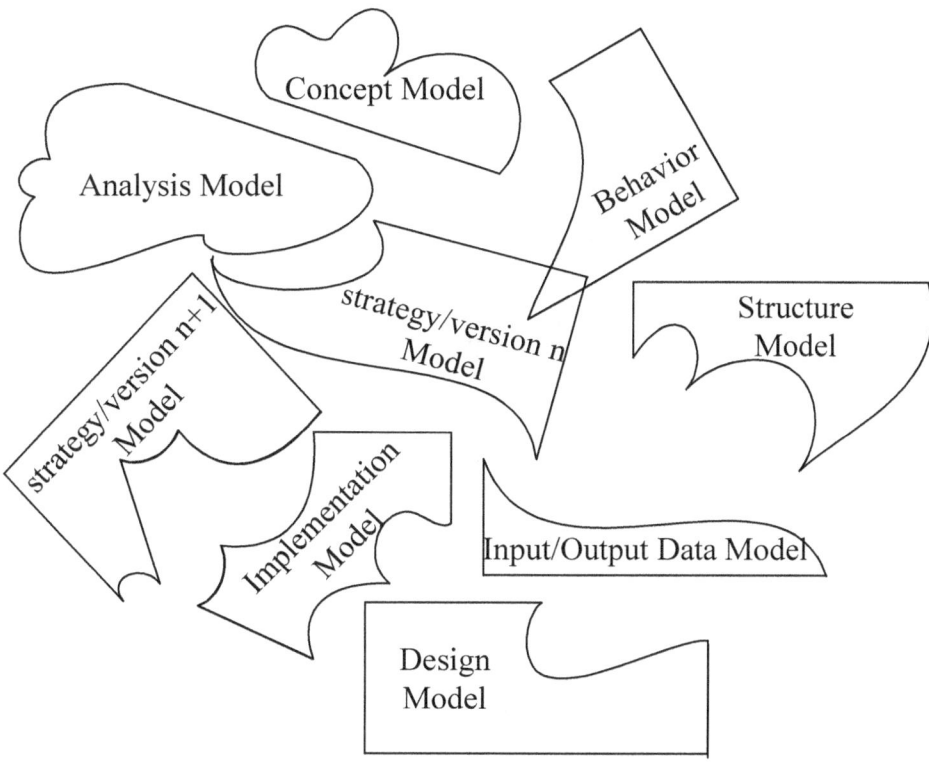

Figure 2-5   The Non-architectural Approach Picks a Model for Each View

The architectural approach, also known as the model singularity approach [Dori95, Dori02, Dori16, Pele02, Sode03], instead of picking many different models, will use only one single coalescence model as shown in Figure 2-6. The strategy/version n, strategy/version n+1, concept, analysis, design, implementation, structure, behavior and input/output data views are all integrated in this multiple views coalescence (MVC) model of systems architecture (SA) [Chao14a, Chao14b, Chao14c, Chao15a, Chao15b, Chao16, Chao17a, Chao17b, Chao17c, Chao17d, Chao17e, Chao17f].

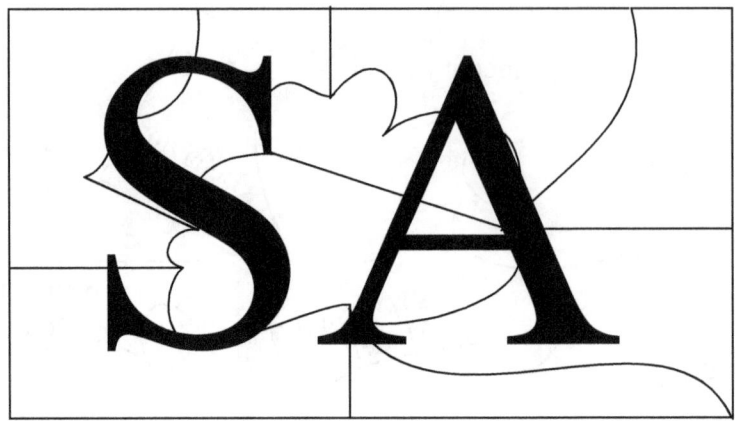

Figure 2-6   Systems Architecture Uses a Coalescence Model

Figure 2-5 has many models. Figure 2-6 has only one model. Comparing Figure 2-5 with Figure 2-6, we unquestionably conclude that an integrated, holistic, united, coordinated, coherent and coalescence model is more favorable than a collection of many heterogeneous and separated models.

2-3 Definition of Systems Architecture

Involved systems are extremely complex in every aspect so that each stakeholder needs a blueprint or model to capture their essential structures and behaviors. Systems architecture is such a blueprint or model.

There are several well-know definitions of systems architecture [Burd10, Craw15, Dam06, Maie09, O'Rou03, Roza11]. ANSI/IEEE 1471-2000 defines systems architecture as: "the fundamental organization of a system, embodied in its components, their relationships to each other and the environment, and the principles governing its designs and evolution." The Open Group defines systems architecture as either "a formal description of a system, or a detailed plan of the system at component level to guide its implementation," or as "the structure of components, their interrelationships, and the principles and guidelines governing their designs and evolution over time" [Rayn09, Toga08].

Concluding the above definitions, we now give systems architecture a definition of our own as shown in Figure 2-7.

> Systems architecture is an integrated whole of a system's multiple views, i.e., structure, behavior and other views, embodied in its assembled components, their interactions with each other and the environment, and the principles and guidelines governing its design and evolution.

Figure 2-7   Definition of Systems Architecture

From the above definition, we find out that systems architecture is an integrated whole of a system's multiple views, i.e., structure, behavior and other views, embodied in its assembled components, their interactions (or handshakes) with each other and the environment, and the principles and guidelines governing its designs and evolution. That is, systems architecture is an integrated and coalescence model of multiple views. In this coalescence model, structure, behavior and other views are all included in it as shown in Figure 2-8. We do not supply each view a respective model in this systems architecture coalescence model.

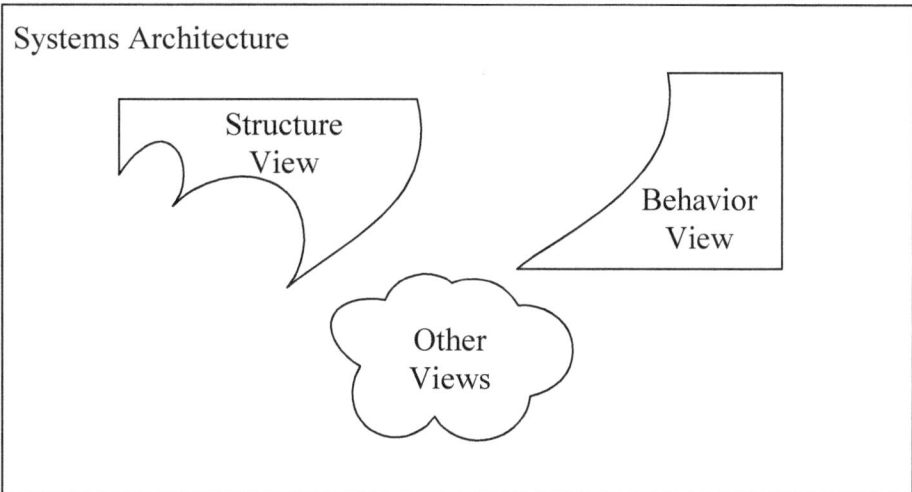

Figure 2-8   All Multiple Views are Included in This Systems Architecture

Since multiple views are embodied in a system's assembled components which belong to the structure view, they shall not exist alone. Multiple views must be loaded on the structure view just like a cargo is loaded on a ship as shown in Figure 2-9. There will be no multiple views if there is no structure view. Stand-alone multiple views are not meaningful.

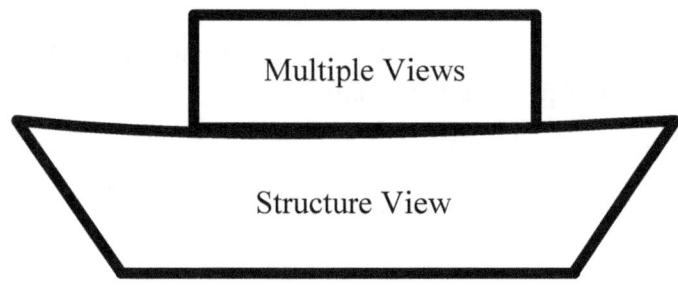

Figure 2-9   Multiple Views Must be Loaded on the Structure View

2-4 Architecture Description Language

An architecture description is a formal description and representation of a system. A description of the systems architecture has to grasp the essence of the system and its details at the same time. In other words, an architecture description not only provides an overall picture that summarizes the whole system, but also contains enough detail that the system can be constructed and validated.

The language for architecture description is called the architecture description language (ADL) [Bass03, Clem02, Clem10, Dike01, Roza11, Shaw96, Tayl09]. An ADL is a special kind of language used in describing the architecture of a system.

Since the architectural approach uses a coalescence model for all multiple views of a system, the foremost duty of ADL is to make the strategy/version n, strategy/version n+1, concept, analysis, designs, implementation, structure, behavior and input/output data views all integrated and coalesced within this architecture description.

2-5 Multiple Views Coalescence to Achieve the Systems Architecture

Systems architecture has been defined as a coalescence model of multiple views. Multiple views coalescence (MVC) uses only a single coalescence model as shown in Figure 2-10. Strategy/version n, strategy/version n+1, concept, analysis, designs, implementation, structure, behavior and input/output data views are all integrated in this MVC architecture.

Figure 2-10   MVC Architecture

Generally, MVC architecture is synonymous with the systems architecture. In other words, multiple views coalescence sets a path to achieve the systems architecture as shown in Figure 2-11.

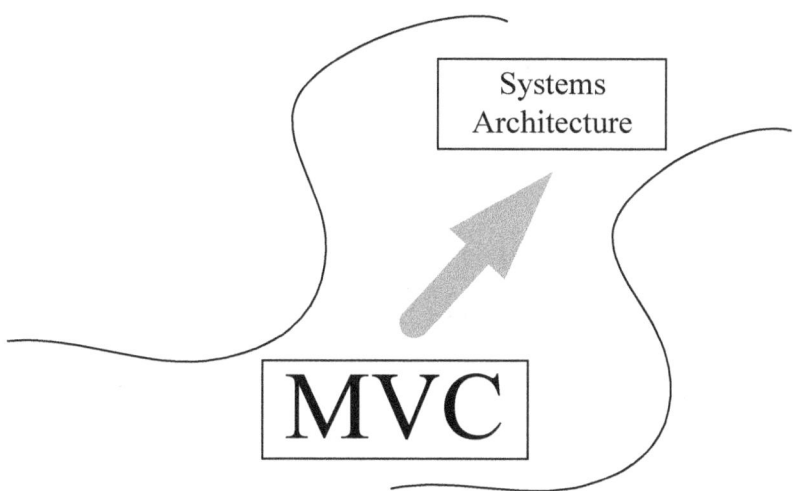

Figure 2-11   MVC to Achieve the Systems Architecture

In the MVC architecture, multiple views must be attached to or built on the systems structure. In other words, multiple views shall not exist alone; they must be loaded on the systems structure just like a cargo is loaded on a ship as shown in Figure 2-12. There will be no multiple views if there is no systems structure. Stand-alone multiple views are not meaningful.

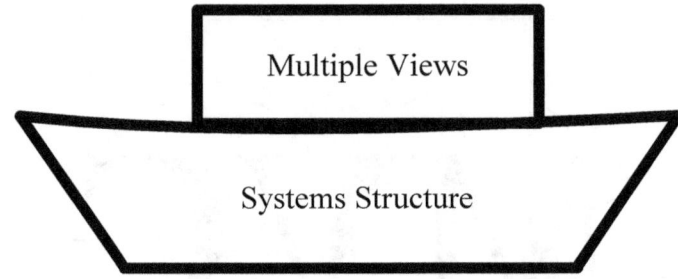

Figure 2-12   Multiple Views Must be Loaded on the Systems Structure

2-6 Integrating the Systems Structures and Systems Behaviors

By integrating the systems structure and systems behavior, we obtain structure-behavior coalescence (SBC) within the system as shown in Figure 2-13.

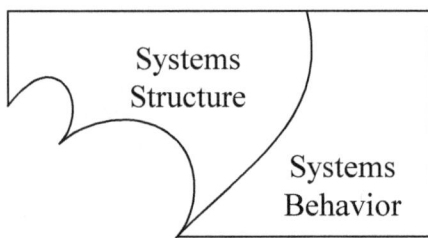

Figure 2-13   Structure-Behavior Coalescence

Structure-behavior coalescence has never been used in any systems model (SM) for systems development except the SBC architecture and object-process methodology (OPM) [Dori95, Dori02, Dori16, Pele02, Sode03]. There are many advantages to use the structure-behavior coalescence approach to integrate the systems structure and systems behavior.

SBC architecture uses a single coalescence model as shown in Figure 2-14. Systems structures and systems behaviors are integrated in this SBC architecture.

Figure 2-14   SBC Architecture

Since systems structures and systems behaviors are so tightly integrated, we sometimes claim that the core theme of SBC architecture is: "Systems Architecture = Systems Structure + Systems Behavior," as shown in Figure 2-15.

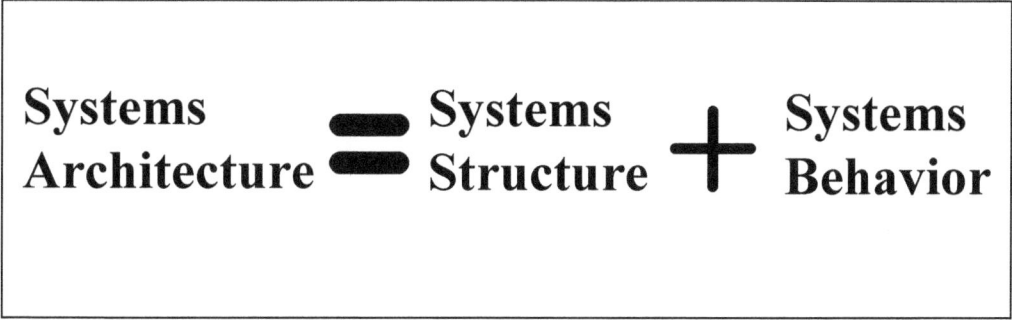

Figure 2-15   Core Theme of SBC Architecture

So far, systems behaviors are separated from systems structures in most cases [Pres09, Somm06]. For example, the well-known structured systems analysis and designs (SSA&D) approach uses structure charts (SC) to represent the systems structure and data flow diagrams (DFD) to represent the systems behavior [Denn08, Kend10, Your99]. SC and DFD are two different models. They are so separated like that there is the "Pacific Ocean" between them, as shown in Figure 2-16.

34

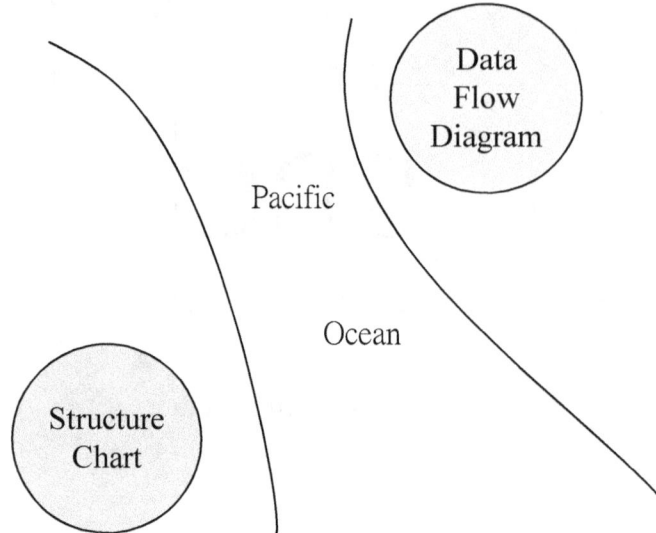

Figure 2-16   Two Heterogeneous and Separated Models

2-7 Structure-Behavior Coalescence to Facilitate Multiple Views Coalescence

Since structure and behavior views are the two most prominent ones among multiple views, integrating the structure and behavior views is clearly the best way to integrate multiple views of a system. In other words, structure-behavior coalescence facilitates multiple views coalescence as shown in Figure 2-17.

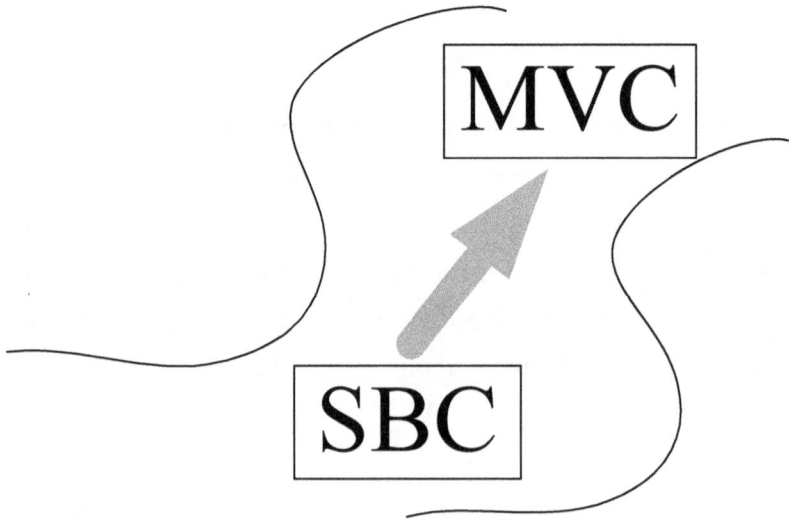

Figure 2-17   SBC Facilitates MVC

2-8 Structure-Behavior Coalescence to Achieve the Systems Architecture

Figure 2-11 declares that multiple views coalescence sets a path to achieve the desired systems architecture with the most efficient approach. Figure 2-17 declares that structure-behavior coalescence facilitates multiple views coalescence.

Combining the above two declarations, we conclude that structure-behavior coalescence sets a path to achieve the systems architecture as shown in Figure 2-18. In this case, SBC architecture is also synonymous with the systems architecture.

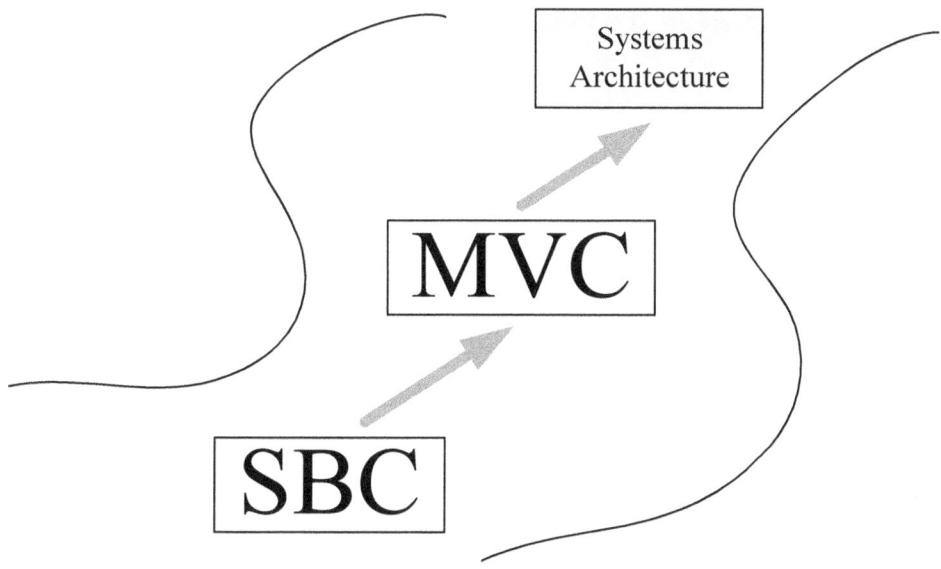

Figure 2-18   SBC to Achieve the Systems Architecture

SBC architecture strongly demands that the structure and behavior views must be coalesced and integrated. This never happens in other architectural approaches such as Zachman Framework [O'Rou03], The Open Group Architecture Framework (TOGAF) [Rayn09, Toga08], Department of Defense Architecture Framework (DoDAF) [Dam06] and Unified Modeling Language (UML) [Rumb91]. Zachman Framework does not offer any mechanism to integrate the structure and behavior views. TOGAF, DoDAF and UML do not, either.

In the SBC architecture, the systems behavior must be attached to or built on the systems structure. In other words, the systems behavior can not exist alone; it must be loaded on the systems structure just like a cargo is loaded on a ship as shown in Figure 2-19. There will be no systems behavior if there is no systems structure. A stand-alone systems behavior is not meaningful.

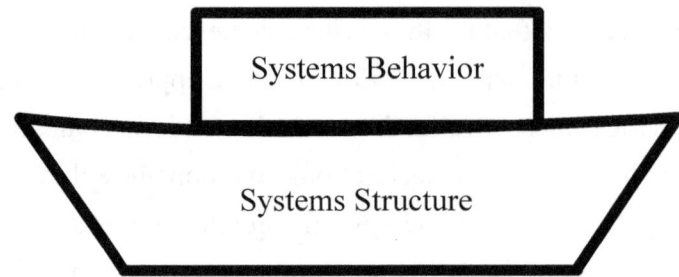

Figure 2-19   Systems Behavior is Loaded on the Systems Structure

2-9 Using SBC-ADL to Construct the Systems Architecture

An architecture description language (ADL) is a special kind of language used in describing the architecture of a system [Shaw96, Tayl09].

A description of the systems architecture has to grasp the essence of a system and its details at the same time. In other words, a systems architecture description not only provides an overall picture that summarizes the system, but also contains enough detail that the system can be constructed and validated.

SBC-ADL uses six fundamental diagrams to describe the integration of systems structure and systems behavior of a system. These diagrams, as shown in Figure 2-20, are: a) architecture hierarchy diagram (AHD), b) framework diagram (FD), c) component operation diagram (COD), d) component connection diagram (CCD), e) structure-behavior coalescence diagram (SBCD) and f) interaction flow diagram (IFD).

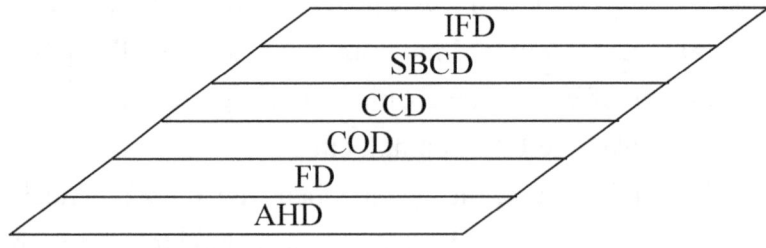

Figure 2-20   Six Fundamental Diagrams of SBC-ADL

SBC-ADL uses AHD, FD, COD, CCD, SBCD and IFD to depict the systems structure and systems behavior of a system as shown in Figure 2-21.

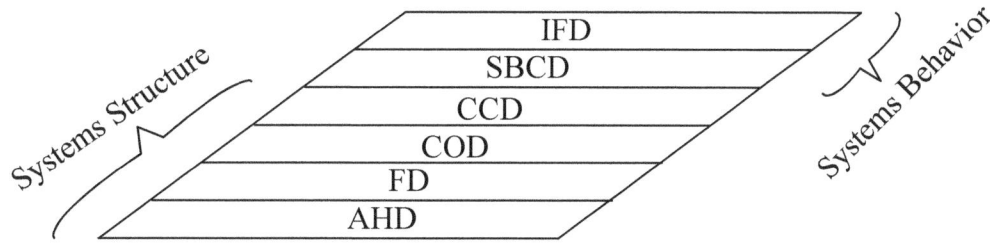

Figure 2-21   Systems Structure and Systems Behavior of a System

Examining the SBC-ADL approach, we find out that it depicts the systems structure first and then depicts the systems behavior later, not the other way around. The reason SBC-ADL does so lies in that the systems behavior must be attached to or built on the systems structure. With the systems structure and attached systems behavior, then, we can smoothly get the systems architecture as shown in Figure 2-22.

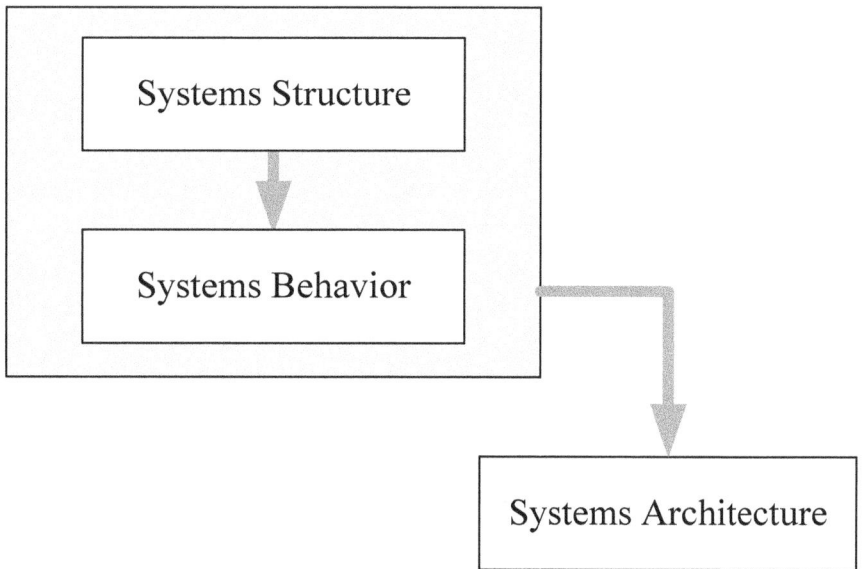

Figure 2-22   Systems Behavior is Attached to the Systems Structure

Let us ask the opposite question. Can the systems structure be attached to or built on the systems behavior? The answer is "No" as shown in Figure 2-23.

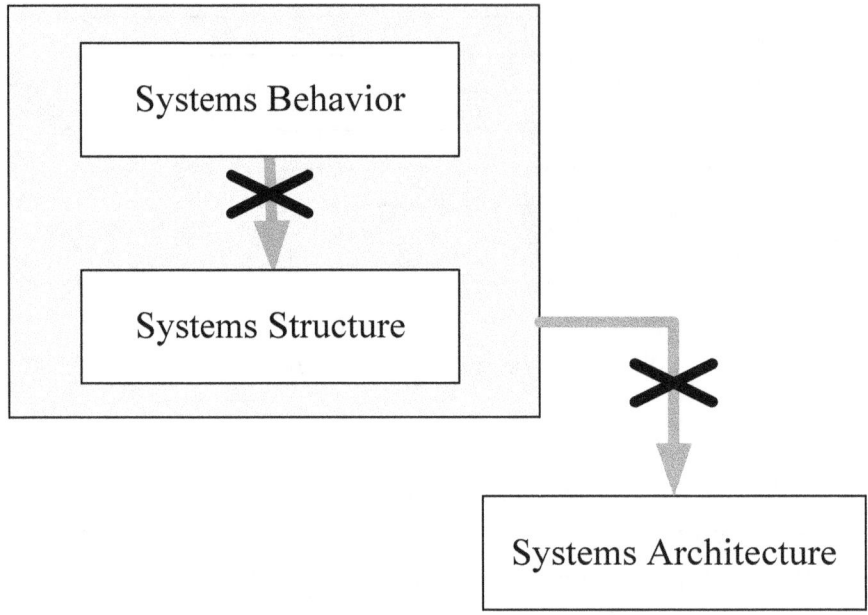

Figure 2-23   Systems Structure is not Attached to the Systems Behavior

In the SBC-ADL, systems behavior must be attached to or built on the systems structure. In other words, the systems behavior shall not exist alone; it must be loaded on the systems structure just like a cargo is loaded on a ship as shown in Figure 2-24. There will be no systems behavior if there is no systems structure. A stand-alone systems behavior is not meaningful.

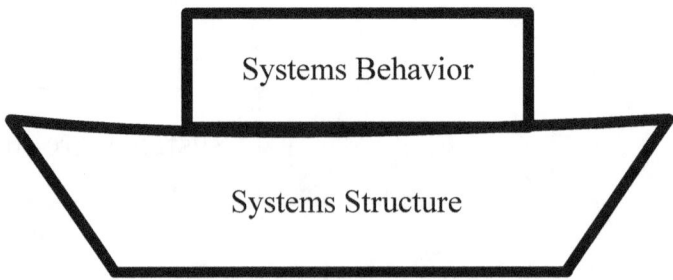

Figure 2-24   Systems Behavior Must be Loaded on the Systems Structure

AHD, FD, COD and CCD belong to systems structure. SBCD and IFD belong to systems behavior. Concluding the above discussion, we perceive that SBC-ADL will describe AHD, FD, COD and CCD first then describe SBCD and IFD later when it constructs the systems architecture of a system.

2-10 SBC Model Singularity

Channel-Based Single-Queue SBC Process Algebra (C-S-SBC-PA) [Chao17a], Channel-Based Multi-Queue SBC Process Algebra (C-M-SBC-PA) [Chao17b], Channel-Based Infinite-Queue SBC Process Algebra (C-I-SBC-PA) [Chao17c], Operation-Based Single-Queue SBC Process Algebra (O-S-SBC-PA) [Chao17d], Operation-Based Multi-Queue SBC Process Algebra (O-M-SBC-PA) [Chao17e] and Operation-Based Infinite-Queue SBC Process Algebra (O-I-SBC-PA) [Chao17f] are the six specialized SBC process algebras. The SBC process algebra (SBC-PA) shown in Figure 2-25 is a model singularity approach.

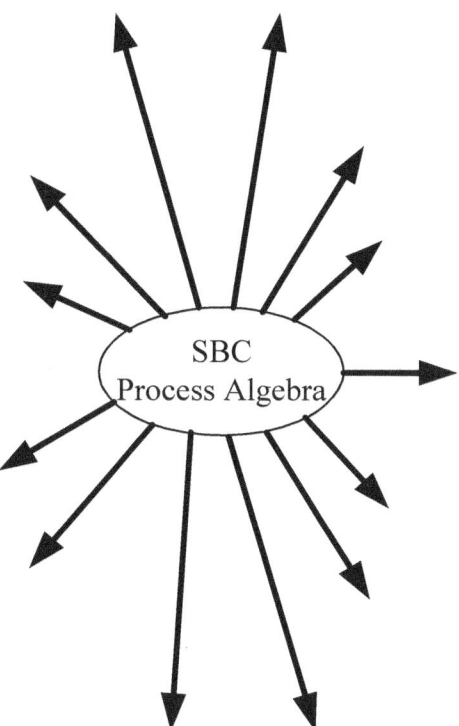

Figure 2-25   SBC-PA is a Model Singularity Approach.

The SBC architecture description language (SBC-ADL) is also a model singularity approach. With SBC mind set sitting in the kernel, the SBC-ADL single model shown in Figure 2-26 is therefore able to represent all structural views such as architecture hierarchy diagram (AHD), framework diagram (FD), component operation diagram (COD), component connection diagram (CCD) and behavioral views such as structure-behavior coalescence diagram (SBCD), interaction flow diagram (IFD).

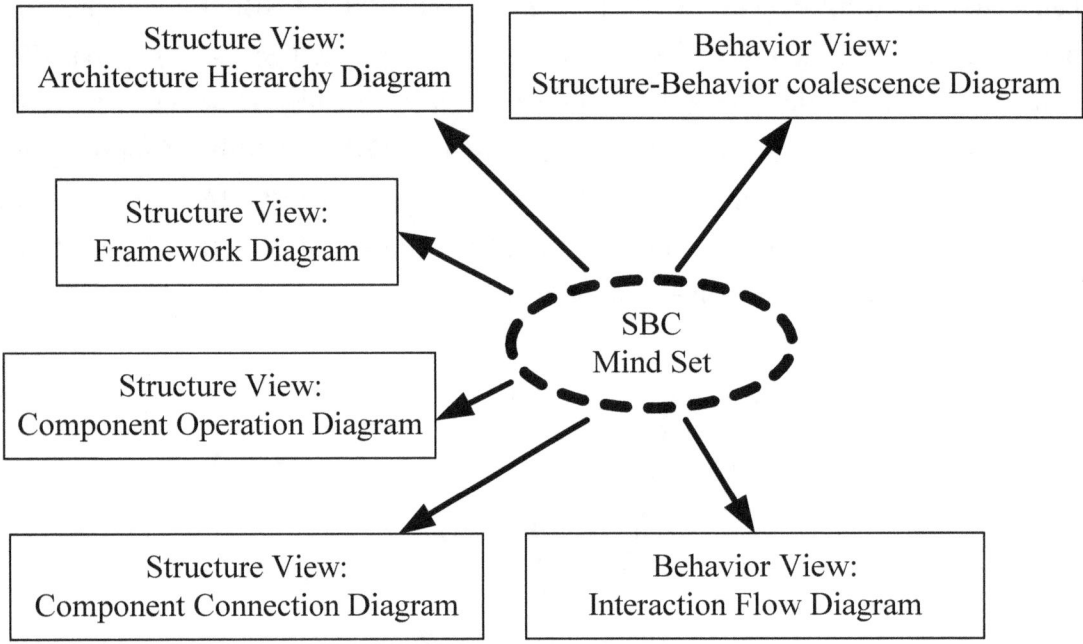

Figure 2-26   SBC-ADL is a Model Singularity Approach.

The combination of SBC process algebra (SBC-PA) and SBC architecture description language (SBC-ADL) is shown in Figure 2-27, again as a model singularity approach.

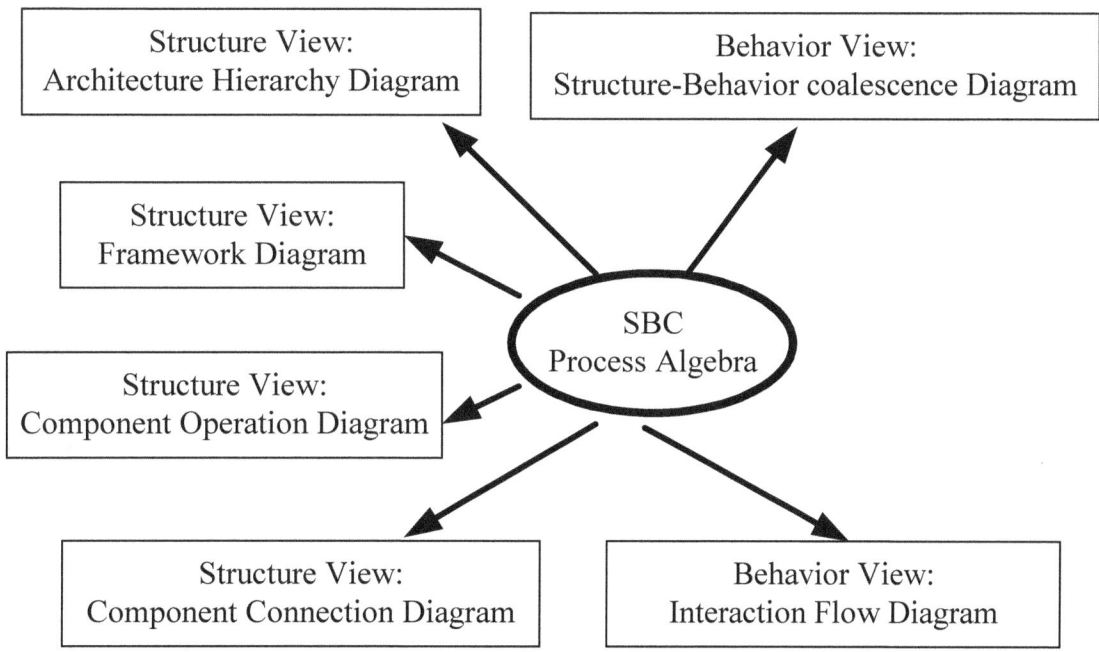

Figure 2-27   SBC Model is a Model Singularity Approach.

PART II: SBC ARCHITECTURE DESCRIPTION LANGUAGE

Chapter 3: Systems Structure

SBC-ADL uses the architecture hierarchy diagram, framework diagram, component operation diagram and component connection diagram to depict the systems structure of a system.

3-1 Architecture Hierarchy Diagram

Systems architects use an architecture hierarchy diagram (AHD) to define the multi-level (hierarchical) decomposition and composition of a system. AHD is the first fundamental diagram to achieve structure-behavior coalescence.

3-1-1 Decomposition and Composition

The following is an example of systems decomposition and composition. The *Computer* system consists of *Monitor*, *Keyboard*, *Mouse* and *Case*, as shown in Figure 3-1. The *Monitor*, *Keyboard*, *Mouse* and *Case* are subsystems comprising the *Computer* system.

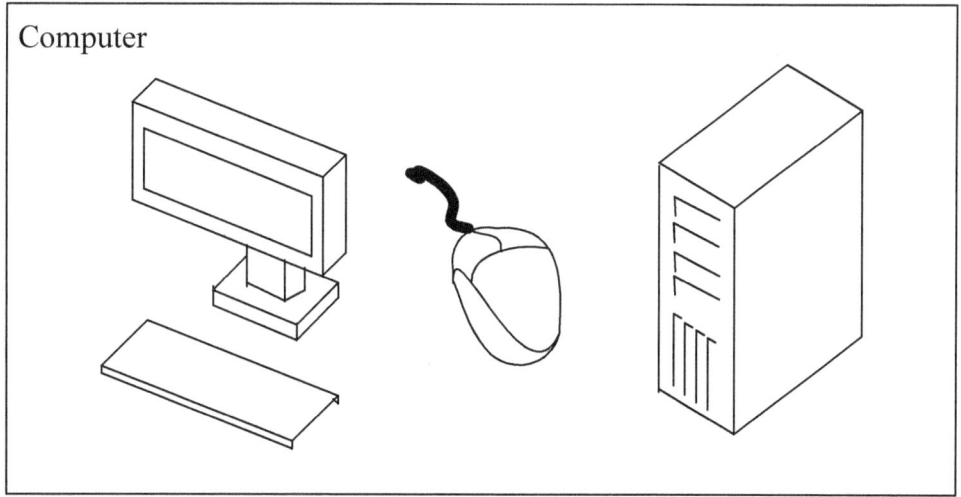

Figure 3-1   Decomposition and Composition of the *Computer* System

Another example indicates that the *Tree* system is composed of *Root* and *Stem*, as shown in Figure 3-2. In this example, we would say that the *Root* and *Stem* are subsystems, respectively, while the *Tree* system consists of its subsystems.

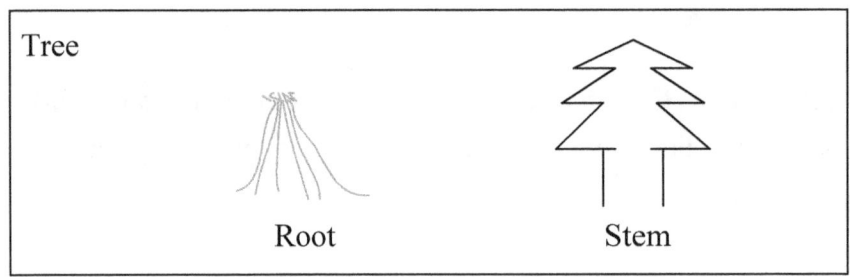

Figure 3-2   Decomposition and Composition of the *Tree* System

The last example demonstrates that the *SBC_Book* system is composed of *Chapter_1*, *Part_1* and *Part_2*, as shown in Figure 3-3. In this example, we would say that *Chapter_1*, *Part_1* and *Part_2* are subsystems, respectively while the *SBC_Book* system consists of its subsystems.

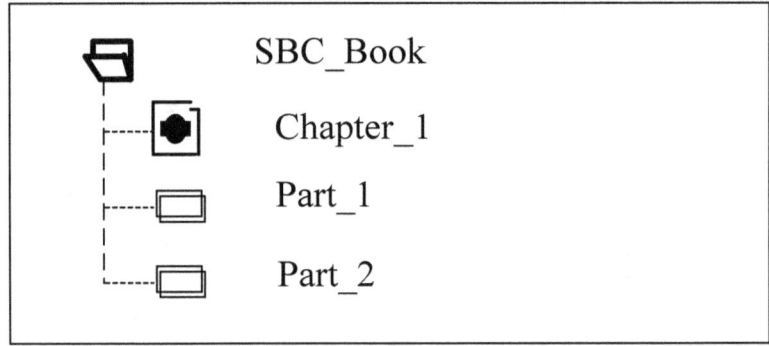

Figure 3-3   Decomposition and Composition of the *SBC_Book* System

Architecture hierarchy diagram (AHD) is used to define the decomposition and composition of a system. As an example, Figure 3-4 shows an AHD of the *Computer* system. We clearly observe that the *Computer* system is composed of *Monitor*, *Keyboard*, *Mouse* and *Case*.

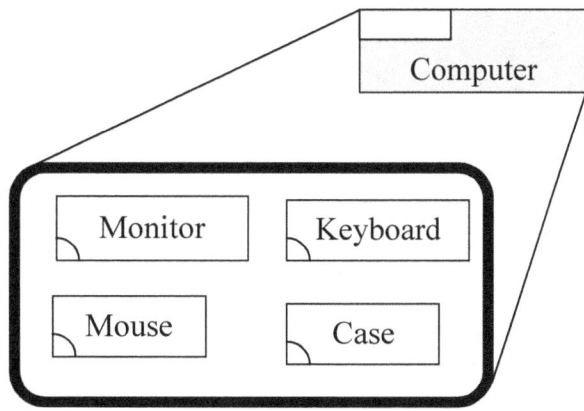

Figure 3-4   AHD of the *Computer* System

As a second example, Figure 3-5 shows an AHD of the *Tree* system. We clearly observe that the *Tree* system is composed of *Root* and *Stem*.

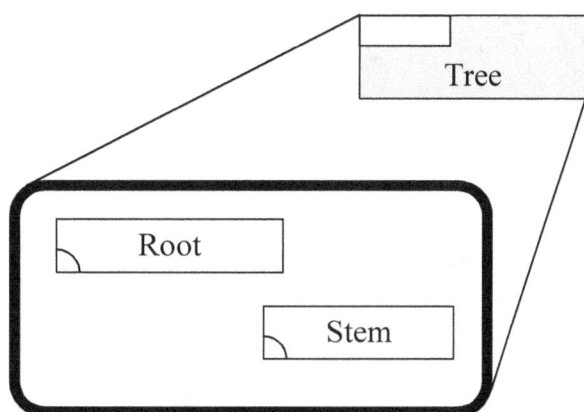

Figure 3-5   AHD of the *Tree* System

As a third example, Figure 3-6 shows an AHD of the *SBC_Book* system. We clearly observe that the *SBC_Book* is composed of *Chapter_1*, *Part_1* and *Part_2*.

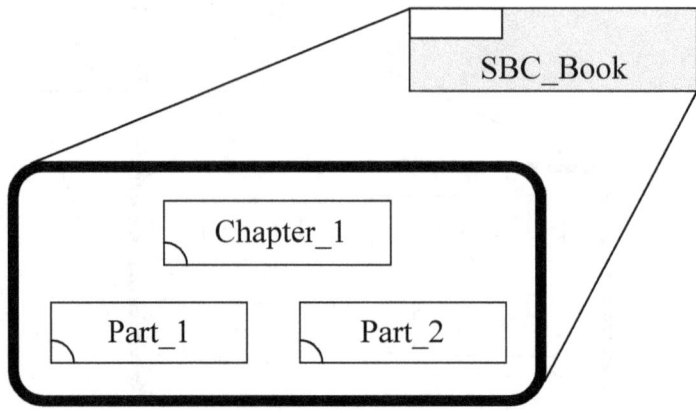

Figure 3-6   AHD of the *SBC_Book* system

3-1-2 Multi-Level Decomposition and Composition

The subsystem may also contain subsystems as we further decompose it. For example, *Case* is a subsystem of the *Computer*, and we can further decompose it into *Motherboard*, *Hard_Disk*, *Power_Supply* and *DVD_Disk*, as shown in Figure 3-7.

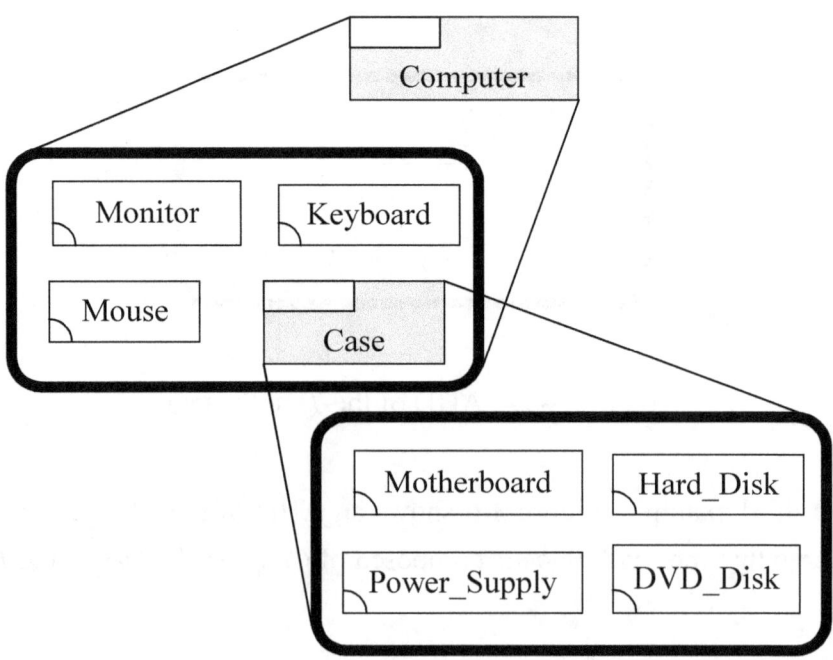

Figure 3-7 Multi-Level Decomposition/Composition of the *Computer* System

As a second example, *Stem* is a subsystem of the *Tree*, and we can further decompose it into *Trunk* and *Leaf*, as shown in Figure 3-8.

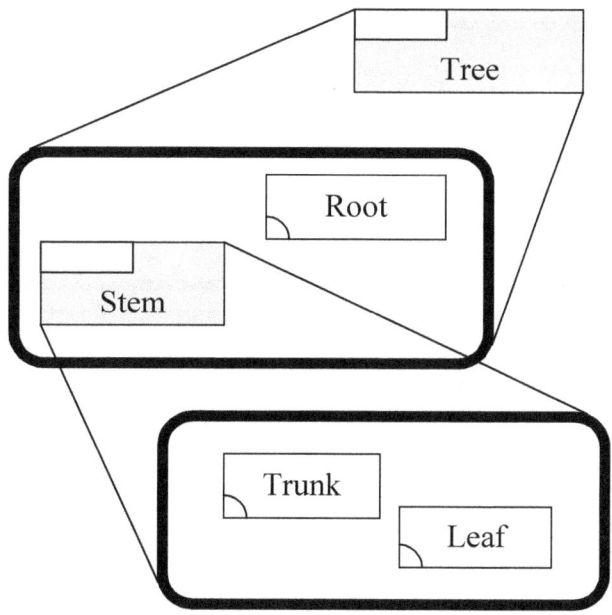

Figure 3-8   Multi-Level Decomposition/Composition of the *Tree* System

As a third example, *Part_1* is a subsystem of the *SBC_Book*, and we can further decompose it into *Chapter_2* and *Chapter_3*; *Part_2* is also a subsystem of the *SBC_Book*, and we can further decompose it into *Chapter_4* and *Chapter_5*, as shown in Figure 3-9.

50

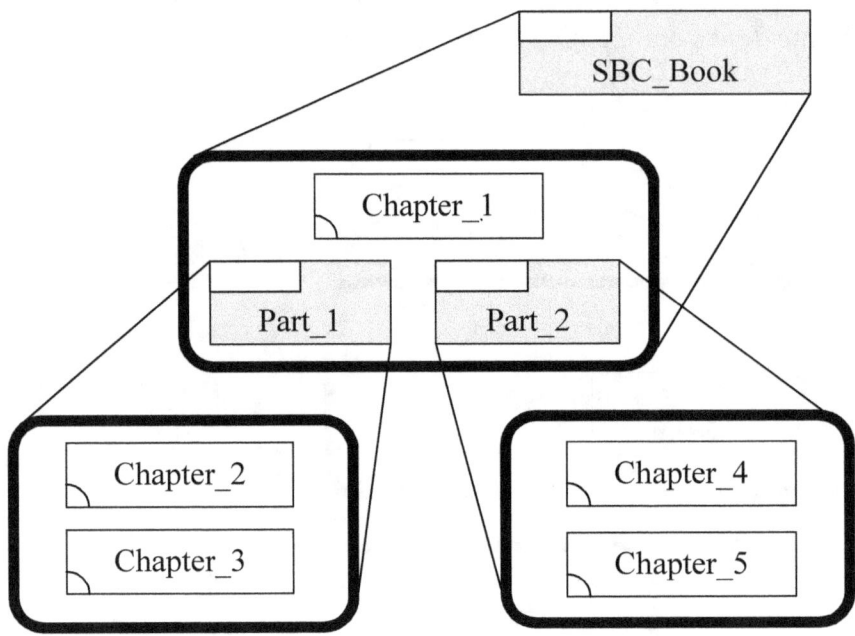

Figure 3-9   Multi-Level Decomposition/Composition of the *SBC_Book* System

Generally speaking, multi-level decomposition and composition of a system is applied often in constructing its architecture. To make a complex system look simple, the mechanism of multi-level composition and decomposition should always be used.

3-1-3 Aggregated and Non-Aggregated Systems

Any subsystem (at any level) involved with multi-level decomposition and composition of a system is either aggregated or non-aggregated. The definition of aggregated and non-aggregated systems is shown in Figure 3-10.

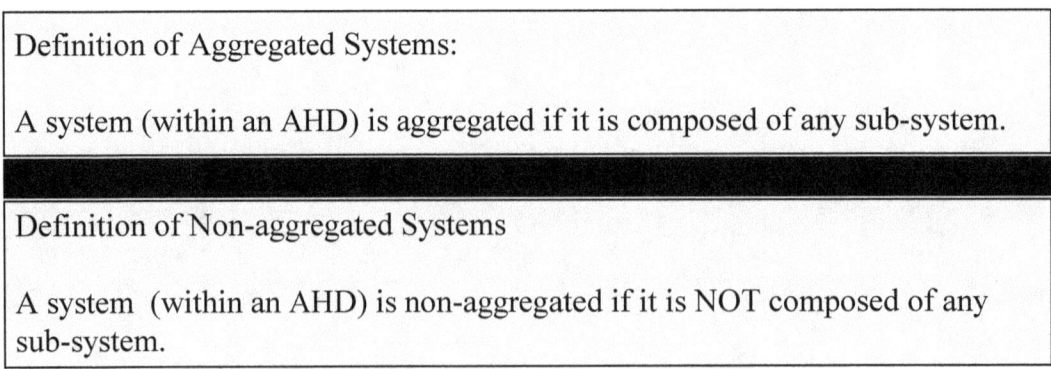

Figure 3-10   Definition of Aggregated and Non-aggregated Systems

Non-aggregated systems are sometimes referred to as components, parts, entities, objects and building blocks [Chao14a, Chao14b, Chao14c].

In the multi-level systems decomposition and composition, any system is either aggregated or non-aggregated, but not both. For example, in Figure 3-4, *Case* is a non-aggregated system, not an aggregated system. As an interesting contrast, in Figure 3-7, *Case* is an aggregated system, not a non-aggregated system.

As a second example, in Figure 3-5, *Stem* is a non-aggregated system, not an aggregated system. As an interesting contrast, in Figure 3-8, *Stem* is an aggregated system, not a non-aggregated system.

As a third example, in Figure 3-6, *Part_1 and Part_2* are non-aggregated systems, not aggregated systems. As an interesting contrast, in Figure 3-9, *Part_1* and *Part_2* are aggregated systems, not non-aggregated systems.

3-2 Framework Diagram

Framework diagram (FD) enables systems architects to examine the multi-layer (also referred to as multi-tier) decomposition and composition of a system. FD is the second fundamental diagram to achieve structure-behavior coalescence.

3-2-1 Multi-Layer Decomposition and Composition

Decomposition and composition of a system can also be represented in a multi-layer (or multi-tier) manner. We draw a framework diagram (FD) for the multi-layer decomposition and composition of a system.

As an example, Figure 3-11 shows a FD of the *Computer* system. In the figure, *Technology_SubLayer_2* contains *Monitor*, *Keyboard* and *Mouse*; *Technology_SubLayer_1* contains *Motherboard*, *Hard_Disk*, *Power_Supply* and *DVD_Disk*.

52

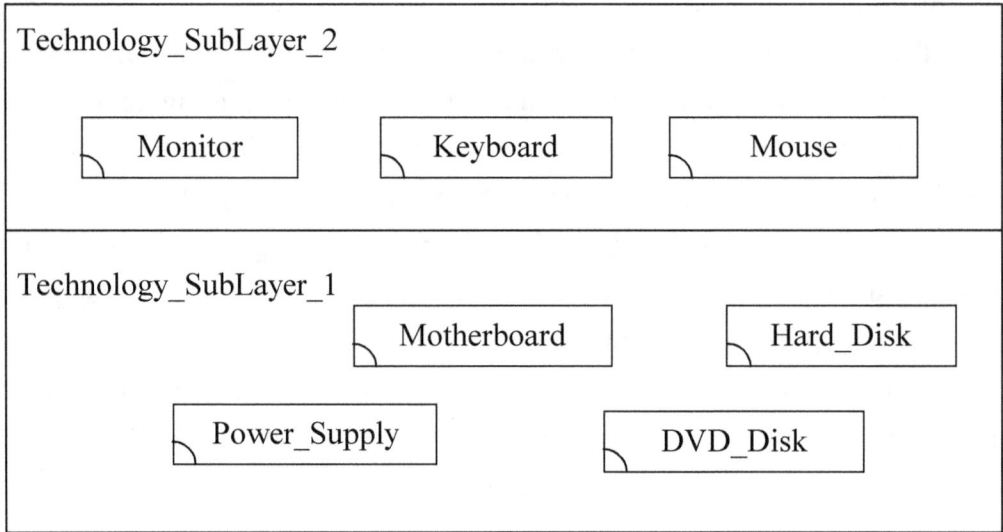

Figure 3-11   FD of the *Computer* System

As a second example, Figure 3-12 shows a FD of the *Tree* system. In the figure, *Technology_SubLayer_2* contains *Root*; *Technology_SubLayer_1* contains *Trunk* and *Leaf*.

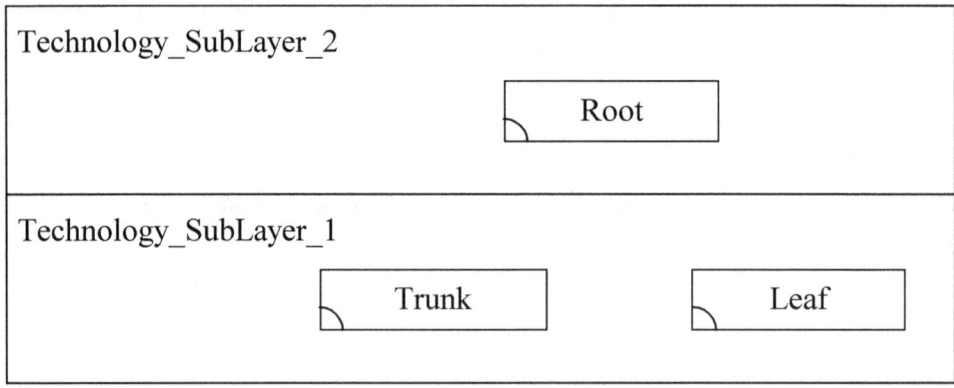

Figure 3-12   FD of the *Tree* System

As a third example, Figure 3-13 shows a FD of the *SBC_Book* system. In the figure, *Technology_SubLayer_2* contains *Chapter_1*; *Technology_SubLayer_1* contains *Chapter_2*, *Chapter_3*, *Chapter_4* and *Chapter_5*.

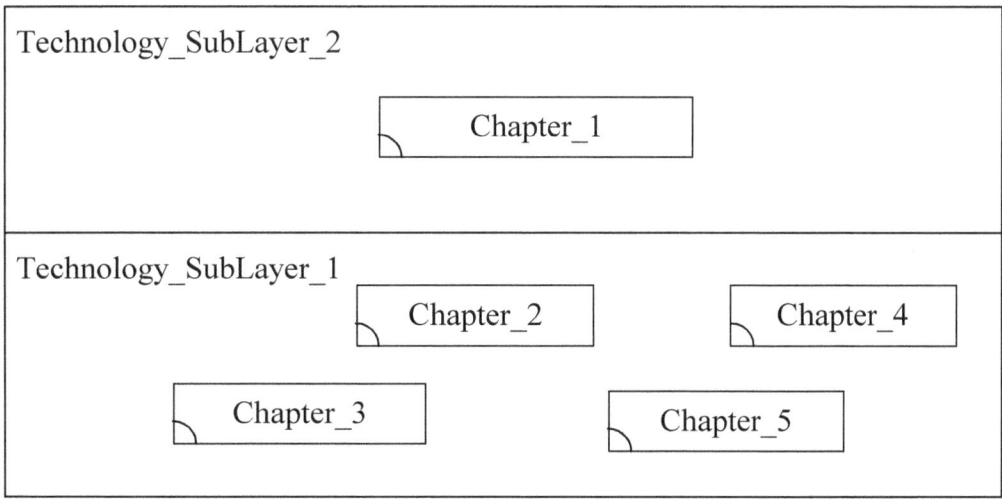

Figure 3-13　FD of the *SBC_Book* System

3-2-2 Only Non-Aggregated Systems Appearing in Framework Diagrams

Both aggregated and non-aggregated systems are displayed in the multi-level AHD decomposition and composition of a system. As an interesting contrast, only non-aggregated systems shall appear in the multi-layer FD decomposition and composition of a system.

For example, Figure 3-7 in the previous section shows an AHD of the *Computer* system in which both aggregated systems such as *Computer*, *Case* and non-aggregated systems such as *Monitor*, *Keyboard*, *Mouse*, *Motherboard*, *Hard_Disk*, *Power_Supply*, *DVD_Disk* are displayed. As an interesting contrast, Figure 3-11 in the previous section shows a FD of the *Computer* system in which only non-aggregated systems such as *Monitor*, *Keyboard*, *Mouse*, *Motherboard*, *Hard_Disk*, *Power_Supply* and *DVD_Disk* are displayed.

For a second example, Figure 3-8 in the previous section shows an AHD of the *Tree* system in which both aggregated systems such as *Tree*, *Stem* and non-aggregated systems such as *Root*, *Trunk*, *Leaf* are displayed. As an interesting contrast, Figure 3-12 in the previous section shows a FD of the *Tree* system in which only non-aggregated systems such as *Root*, *Trunk* and *Leaf* are displayed.

For a third example, Figure 3-9 in the previous section shows an AHD of the *SBC_Book* system in which both aggregated systems such as *SBC_Book*, *Part_1*, *Part_2* and non-aggregated systems such as *Chapter_1*, *Chapter_2*, *Chapter_3*, *Chapter_4*, *Chapter_5* are displayed. As an interesting contrast, Figure 3-13 in the previous section shows a FD of the *SBC_Book* system in which only non-aggregated systems such as *Chapter_1*, *Chapter_2*, *Chapter_3*, *Chapter_4* and *Chapter_5* are displayed.

3-3 Component Operation Diagram

Systems architects use a component operation diagram (COD) to display all components' operations of a system. COD is the third fundamental diagram to achieve structure-behavior coalescence.

3-3-1 Operations of Components

An operation provided by each component represents a procedure, or method, or function of the component. If other systems request this component to perform an operation, then shall use it to accomplish the operation request.

Each component in a system must possess at least one operation. A component should not exist in a system if it does not possess any operation. Figure 3-14 shows that component *SalePurchase_UI* has four operations: *SaleInputClick*, *SalePrintClick*, *PurchaseInputClick* and *PurchasePrintClick*.

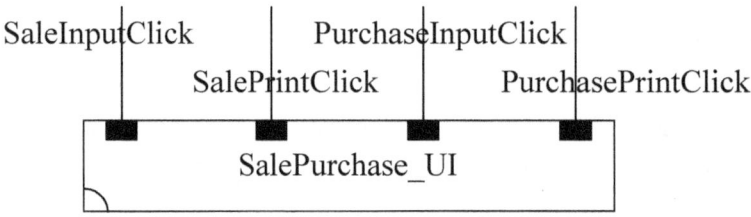

Figure 3-14   Four Operations of the *SalePurchase_UI* Component

An operation formula is utilized to fully represent an operation. An operation formula includes a) operation name, b) input parameters and c) output parameters as shown in Figure 3-15.

Operation_Name (In i_1, i_2, ..., i_m ; Out o_1 , o_2, ..., o_n)

Figure 3-15   Operation Formula

Operation name is the name of this operation. In a system, every operation name should be unique. Duplicate operation names shall not be allowed in any system.

An operation may have several input and output parameters. The input and output parameters, gathered from all operations, represent the input data and output data views of a system [Date03, Elma10]. As shown in Figure 3-16, component *SalePrint_UI* possesses the *ShowModal* operation which has no input/output parameter; component *SalePrint_UI* also possesses the *SalePrintButtonClick* operation which has the *sDate* and *sNo* input parameters (with the arrow direction pointing to the component) and the *s_report* output parameter (with the arrow direction opposite to the component).

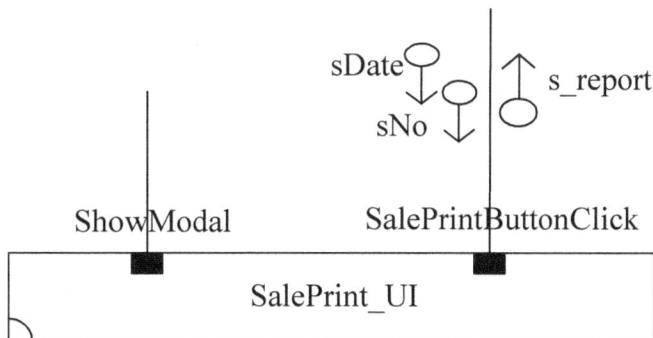

Figure 3-16   Input/Output Parameters of *SalePrintButtonClick*

Data formats of input and output parameters can be described by data type specifications. There are two sets of data types: primitive and composite [Date03, Elma10]. Figure 3-17 shows the primitive data type specification of the *sDate* and *sNo* input parameters occurring in the *SalePrintButtonClick(In sDate, sNo; Out s_report)* operation formula.

Parameter	Data Type	Instances
sDate	Text	20100517, 20100612
sNo	Text	001, 002

Figure 3-17   Primitive Data Type Specification

Figure 3-18 shows the composite data type specification of the *s_report* output parameter occurring in the *SalePrintButtonClick(In sDate, sNo; Out s_report)* operation formula.

Parameter	s_report
Data Type	TABLE of Sale Date : Text Sale No : Text Customer : Text ProductNo : Text Quantity : Integer UnitPrice : Real Total : Real End TABLE;
Instances	Sale Date : 20100517 Sale No : 001 Customer : Larry Fink ProductNo / Quantity / UnitPrice A12345 / 400 / 100.00 A00001 / 300 / 200.00 Total : 100,000.00

Figure 3-18   Composite Data Type Specification

3-3-2 Drawing the Component Operation Diagram

For a system, COD is used to display all components' operations. Figure 3-19 shows the *Multi-Tier Personal Data System's COD*. In the figure, component *MTPDS_UI* has two operations: *Calculate_AgeClick* and *Calculate_OverweightClick*; component *Age_Logic* has one operation: *Calculate_Age*; component *Overweight_Logic* has one operation: *Calculate_Overweight*; component *Personal_Database* has two operations: *Sql_DateOfBirth_Select* and *Sql_SexHeightWeight_Select*.

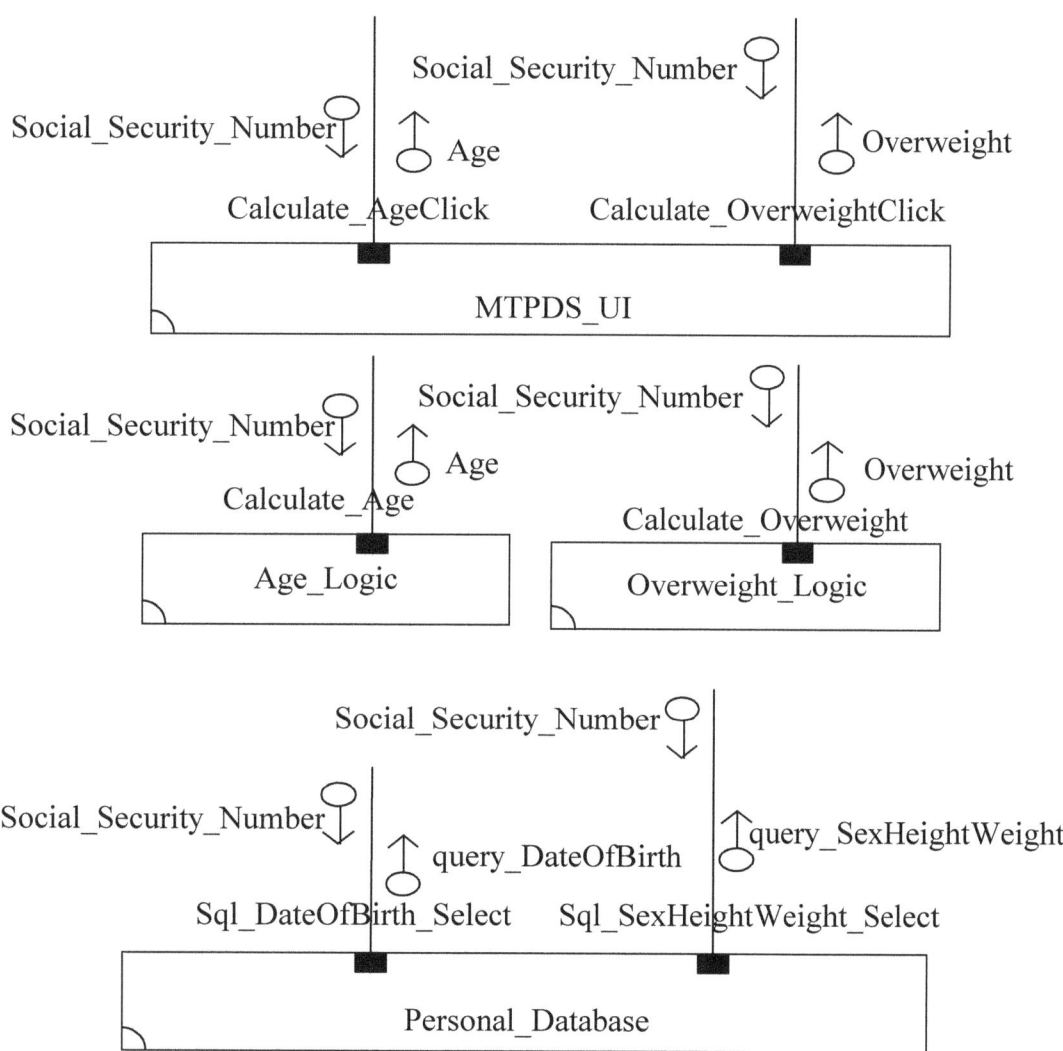

Figure 3-19   COD of the *Multi-Tier Personal Data System*

The operation formula of *Calculate_AgeClick* is *Calculate_AgeClick(In Social_Security_Number; Out Age)*. The operation formula of *Calculate_OverweightClick* is *Calculate_OverweightClick(In Social_Security_Number; Out Overweight)*. The operation formula of *Calculate_Age* is *Calculate_Age(In Social_Security_Number; Out Age)*. The operation formula of *Calculate_Overweight* is *Calculate_Overweight(In Social_Security_Number; Out Overweight)*. The operation formula of *Sql_DateOfBirth_Select* is *Sql_DateOfBirth_Select(In Social_Security_Number; Out query_DateOfBirth)*. The operation formula of *Sql_SexHeightWeight_Select* is *Sql_SexHeightWeight_Select(In Social_Security_Number; Out query_SexHeightWeight)*.

Figure 3-20 shows the primitive data type specification of the *Social_Security_Number* input parameter and the *Age*, *Overweight* output parameters.

Parameter	Data Type	Instances
Social_Security_Number	Text	424-87-3651, 512-24-3722
Age	Integer	28, 56
Overweight	Boolean	Yes, No

Figure 3-20   Primitive Data Type Specification

Figure 3-21 shows the composite data type specification of the *query_DateOfBirth* output parameter occurring in the *Sql_DateOfBirth_Select(In Social_Security_Number; Out query_DateOfBirth)* operation formula.

Parameter	*query_DateOfBirth*

Data Type	TABLE of
	  Social_Security_Number : Text
	  Age : Integer
	End TABLE ;

Instances	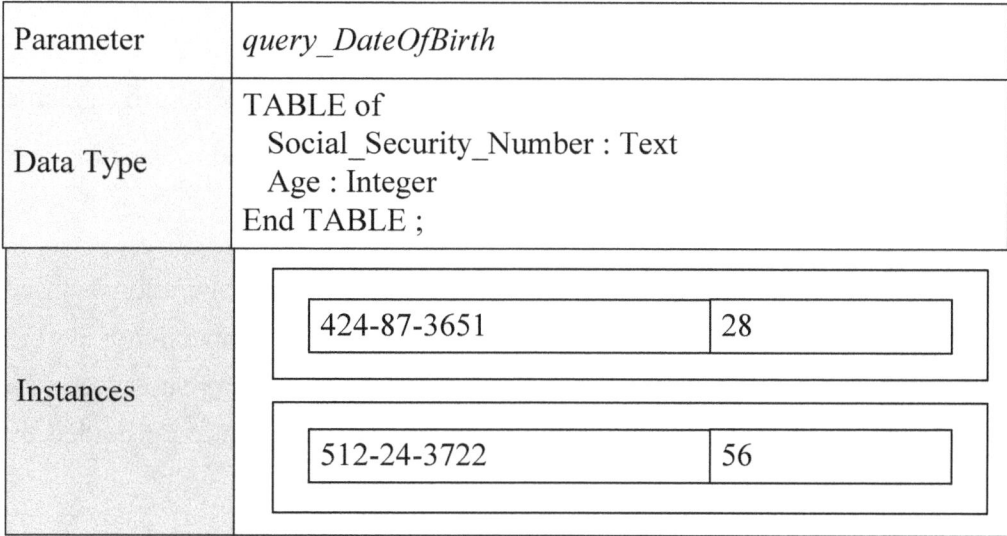

Figure 3-21   Composite Data Type Specification

Figure 3-22 shows the composite data type specification of the *query_SexHeightWeight* output parameter occurring in the *Sql_SexHeightWeight_Select(In Social_Security_Number; Out query_SexHeightWeight)* operation formula.

Parameter	*query_SexHeightWeight*

Data Type	TABLE of
	  Social_Security_Number : Text
	  Sex : Text
	  Height : Number
	  Weight : Number
	End TABLE ;

Instances	424-87-3651 Female 162 76 512-24-3722 Male 180 80

Figure 3-22   Composite Data Type Specification

3-4 Component Connection Diagram

A component connection diagram (CCD) is utilized to describe how all components and actors are connected within a system. CCD is the fourth fundamental diagram to achieve structure-behavior coalescence.

3-4-1 Essence of a Connection

A connection implies an operation request. When an operation is used by another subsystem then a connection appears. Accordingly, a connection is defined as the linkage that is constructed when an operation is used by another subsystem. Figure 3-23 shows that Subsystem_A uses the *Salary_Calculation* operation provided by the *Component_B* component.

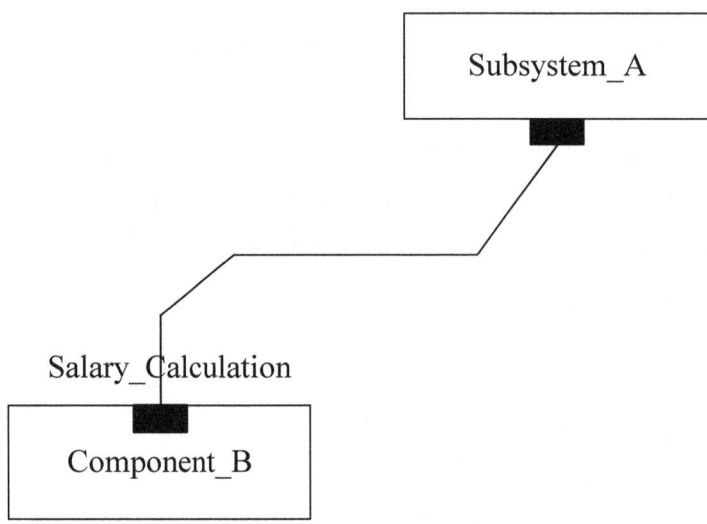

Figure 3-23   A Connection Appears When an Operation is Used

The above figure describes, sufficiently, the essence of a connection. However, we seldom use this kind of drawing. Instead, a simplified drawing of the above figure is often used as shown in Figure 3-24.

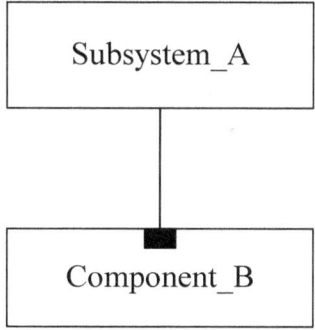

Figure 3-24   Simplified Drawing of a Connection

Since an operation is always provided by a component, there is no doubt that the *Component_B* operation provider is a component. On the contrary, the *Subsystem_A* operation user can be either a component (e.g., *Component_A*) or an actor (e.g., *Actor_A*) as shown in Figure 3-25. An actor belongs to the external environment of a system.

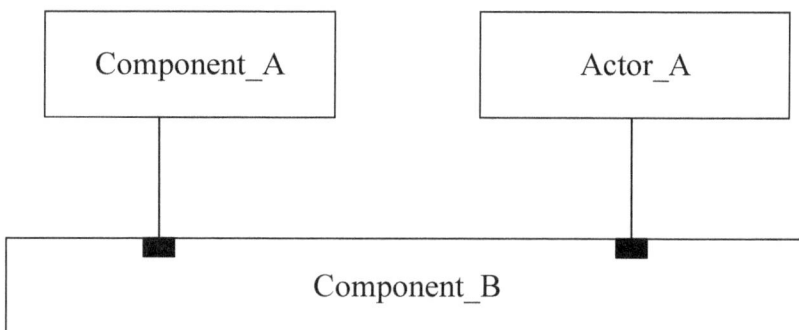

Figure 3-25   Operation User is Either a Component Or an Actor

Within a connection the subsystem (either a component or an actor) using the operation is always entitled the *Client* and the component which provides the operation is always entitled the *Server* as Figure 3-26 shows.

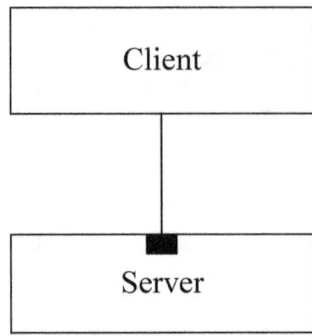

Figure 3-26   Roles of Client and Server Within a Connection

3-4-2 Drawing the Component Connection Diagram

A component connection diagram (CCD) is utilized to describe how all components and actors (in the external environment) are connected within a system. Figure 3-27 exhibits the *Multi-Tier Personal Data System's* COD.

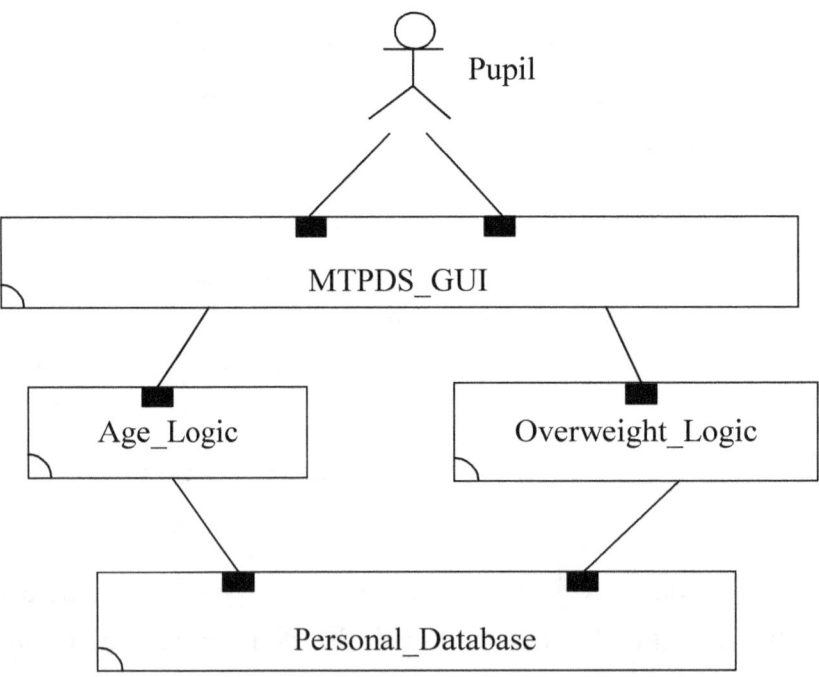

Figure 3-27   CCD of the *Multi-Tier Personal Data System*

In Figure 3-27, actor *Pupil* has two connections with the *MTPDS_UI* component; component *MTPDS_UI* has one connection with each of the *Age_Logic* and *Overweight_Logic* components; component *Age_Logic* has a connection with the *Personal_Database* component; component *Overweight_Logic* has a connection with the *Personal_Database* component.

After finishing the CCD, the formation pattern of the *Multi-Tier Personal Data System* will be constructed; thus the systems structure of the *Multi-Tier Personal Data System* becomes more transparent.

Chapter 4: Systems Behavior

SBC-ADL uses the structure-behavior coalescence diagram and interaction flow diagram to delineate the systems behavior of a system.

4-1 Structure-Behavior Coalescence Diagram

Structure-behavior coalescence diagram (SBCD) enables a system architect to observe the structure and behavior coexisting in a system. SBCD is the fifth fundamental diagram to achieve structure-behavior coalescence.

4-1-1 Purpose of Structure-Behavior Coalescence Diagram

The major aim of the SBC-ADL approach is to achieve the integration of systems structure and systems behavior within a system. SBCD enables a system architect to observe the systems structure and systems behavior coexisting in a system. This is the purpose of utilizing SBCD when architecting the systems architecture.

Figure 4-1 exhibits the *Multi-Tier Personal Data System*'s SBCD In this example, interactions among the *Pupil* actor and the *MTPDS_UI*, *Age_Logic*, *Overweight_Logic* and *Personal_Database* components shall draw forth the *AgeCalculation* and *OverweightCalculation* behaviors.

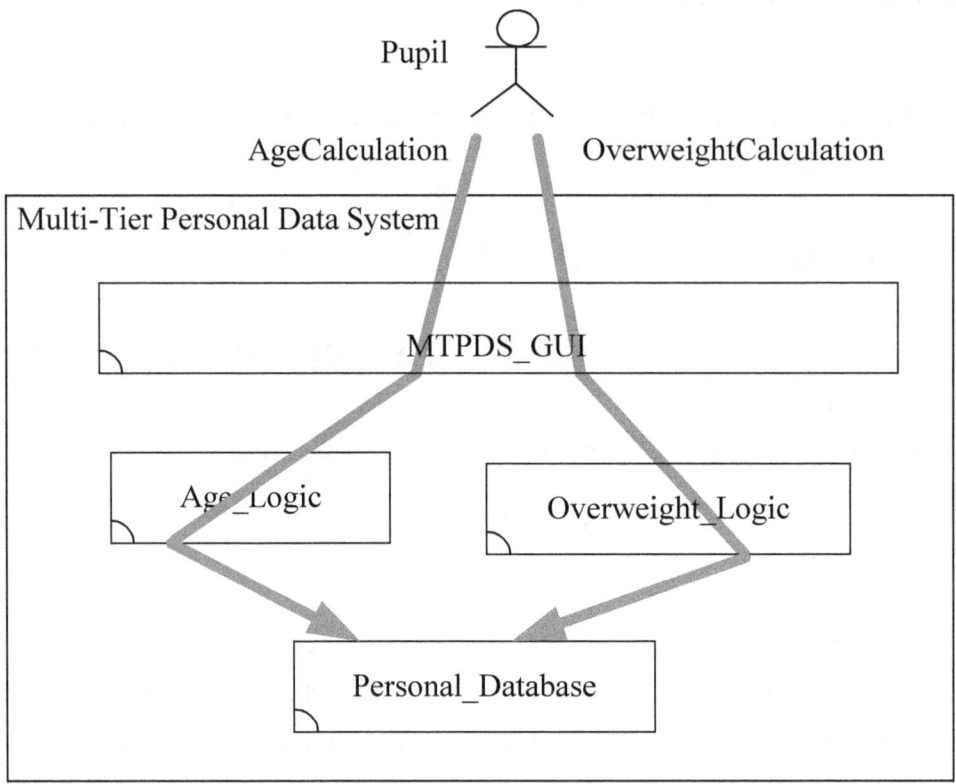

Figure 4-1   SBCD of the *Multi-Tier Personal Data System*

The overall behavior of a system is the aggregation of all its individual behaviors. All individual behaviors are mutually independent of each other. They tend to be executed concurrently [Hoar85, Miln89, Miln99]. For example, the overall behavior of the *Multi-Tier Personal Data System* includes the *AgeCalculation* and *OverweightCalculation* behaviors. In other words, the *AgeCalculation* and *OverweightCalculation* behaviors are combined to produce the overall behavior of the *Multi-Tier Personal Data System*.

The major purpose of using the architectural approach, instead of separating the structure model from the behavior model, is to achieve a coalesced model. In Figure 4-1, systems architects are able to see the systems structure and systems behavior coexisting in a SBCD. That is, in the *Multi-Tier Personal Data System's* SBCD, we not only see its systems structure but also see (at the same time) its systems behavior.

4-1-2 Drawing the Structure-Behavior Coalescence Diagram

Let us now explain the usage of SBCD by constructing a SBCD step by step. The goal of having a SBCD is enabling systems architects to see both the structure and behavior, simultaneously. In order to achieve this goal, a SBCD is drawn by first constructing all of the components, then describing the external environment's actors, and finally describing the interactions among these components and the external environment's actors.

For example, the *Multi-Tier Personal Data System* has two behaviors: *AgeCalculation* and *OverweightCalculation*. After constructing the *Multi-Tier Personal Data System* with all its components, the external environment's actors and the *AgeCalculation* behavior, we obtain the graphical representation as shown in Figure 4-2. In this Figure, the *AgeCalculation* behavior indicates that actor *Pupil* interacts with the *MTPDS_UI* component first, then component *MTPDS_UI* interacts with the *Age_Logic* component later, then component *Age_Logic* interacts with the *Personal_Database* component finally.

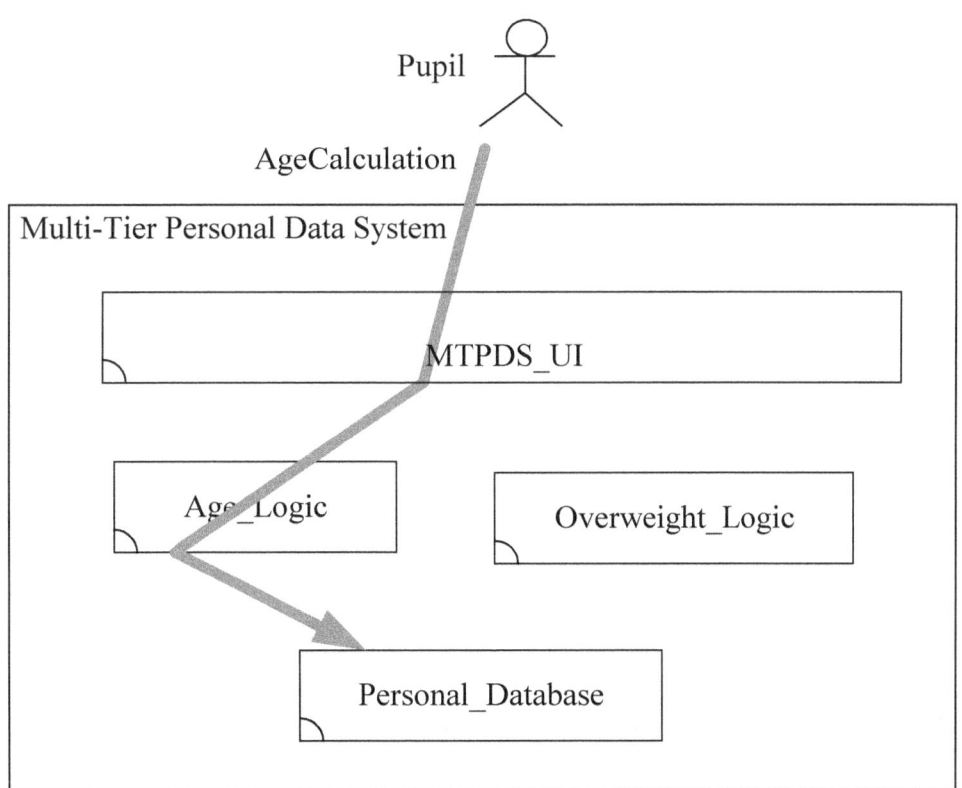

Figure 4-2   All Components, Actors, and the *AgeCalculation* Behavior

Adding the *OverweightCalculation* behavior to Figure 4-2, we then obtain the graphical representation shown in Figure 4-3. In this Figure, the *OverweightCalculation* behavior indicates that actor *Pupil* interacts with the *MTPDS_UI* component first, then component *MTPDS_UI* interacts with the *Overweight_Logic* component later, then component *Overweight_Logic* interacts with the *Personal_Database* component finally.

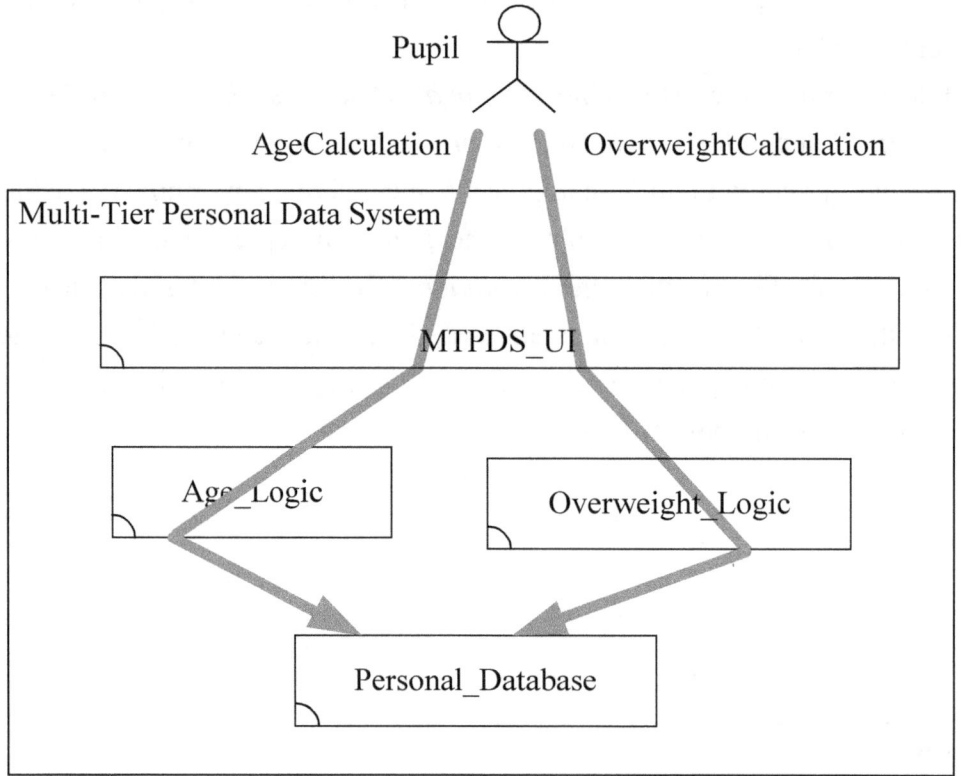

Figure 4-3   Adding the *OverweightCalculation* Behavior to Figure 4-2

After finishing Figure 4-3, we actually have accomplished all the works needed to draw an entire SBCD of the *Multi-Tier Personal Data System*. As a matter of fact, Figure 4-3 shows exactly the *Multi-Tier Personal Data System*'s SBCD.

4-2 Interaction Flow Diagram

An interaction flow diagram (IFD) is utilized to describe each individual behavior of the overall behavior of a system. IFD is the sixth fundamental diagram to achieve structure-behavior coalescence.

4-2-1 Individual Systems Behavior Represented by Interaction Flow Diagram

The overall behavior of a system consists of many individual behaviors. Each individual behavior represents an execution path. An IFD is utilized to represent such an individual behavior.

Figure 4-4 demonstrates that the *Multi-Tier Personal Data System* has two behaviors; thus, it has two IFDs.

Enterprise	IFD
Multi-Tier Personal Data System	AgeCalculation
	OverweightCalculation

Figure 4-4   *Multi-Tier Personal Data System* has Two IFDs

4-2-2 Drawing the Interaction Flow Diagram

Let us now explain the usage of interaction flow diagram (IFD) by drawing an IFD step by step. Figure 4-5 demonstrates an IFD of the *SaleInput* behavior. The X-axis direction is from the left side to right side and the Y-axis direction is from the above to the below. Inside an IFD, there are four elements: a) external environment's actor, b) components, c) interactions and d) input/output parameters. Participants of the interaction, such as the external environment's actor and each component, are laid aside along the X-axis direction on the top of the diagram. The external environment's actor which initiates the sequential interactions is always placed on the most left side of the X-axis. Then, interactions among the external environment's actor and components successively in turn decorate along the Y-axis direction. The first interaction is placed on the top of the Y-axis position. The last interaction is placed on the bottom of the Y-axis position. Each interaction may carry several input and/or output parameters.

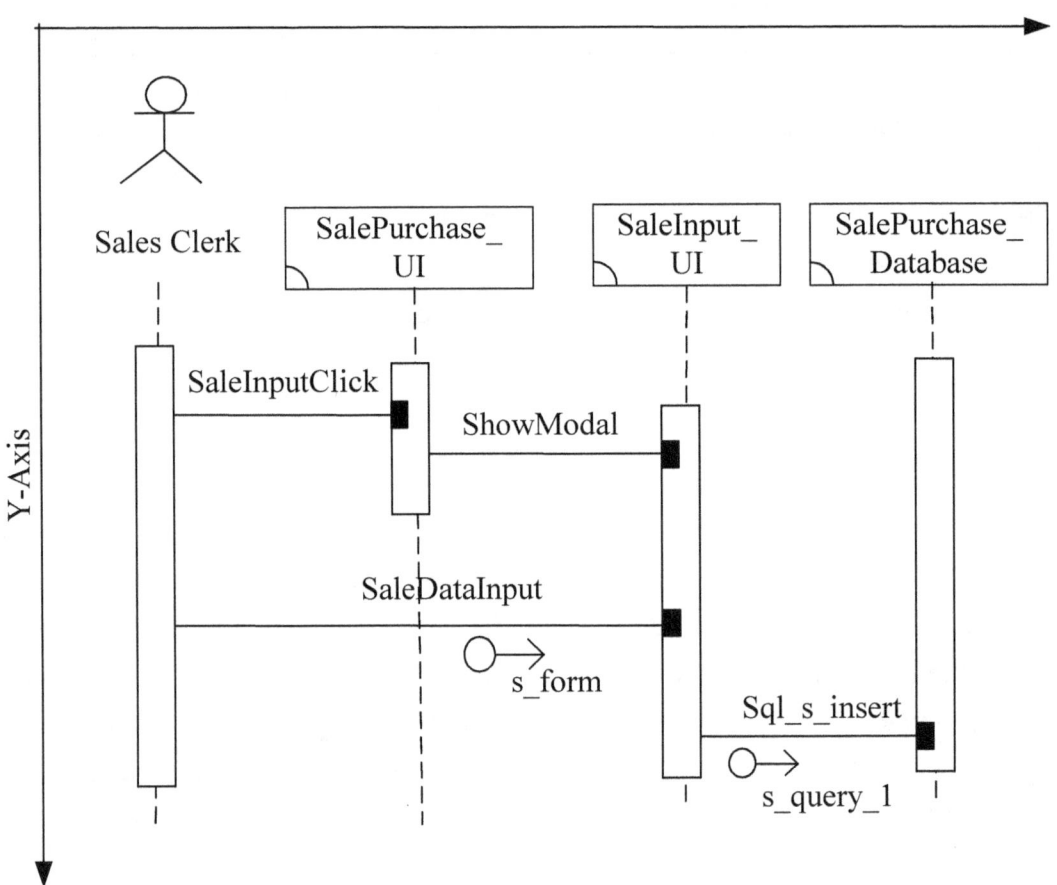

Figure 4-5   IFD of the *SaleInput* Behavior

In Figure 4-5, *Sales Clerk* is an external environment's actor. *SalePurchase_UI*, *SaleInput_UI* and *SalePurchase_Database* are components. *SaleInputClick* is an operation which is provided by the *SalePurchase_UI* component. *ShowModal* is an operation which is provided by the *SaleInput_UI* component. *SaleDataInput* is an operation, carrying the *s_form* input parameter, which is also provided by the *SaleInput_UI* component. *Sql_s_insert* is an operation, carrying the *s_query_1* input parameter, which is provided by the *SalePurchase_Database* component.

The execution path of Figure 4-5 is as follows. First, actor *Sales Clerk* interacts with the *SalePurchase_UI* component through the *SaleInputClick* operation call interaction. Next, component *SalePurchase_UI* interacts with the *SaleInput_UI* component through the *ShowModal* operation call interaction. Continuingly, actor *Sales Clerk* interacts with the *SaleInput_UI* component through the *SaleDataInput* operation call interaction, carrying the *s_form* input parameter. Finally, component *SaleInput_UI* interacts with the *SalePurchase_Database* component through the *Sql_s_insert* operation call interaction, carrying the *s_query_1* input parameter.

For each interaction, the solid line stands for operation call while the dashed line stands for operation return. The operation call and operation return interactions, if using the same operation name, belong to the identical operation. Figure 4-6 exhibits two interactions (operation call interaction and operation return interaction) having the identical "*Request*" operation.

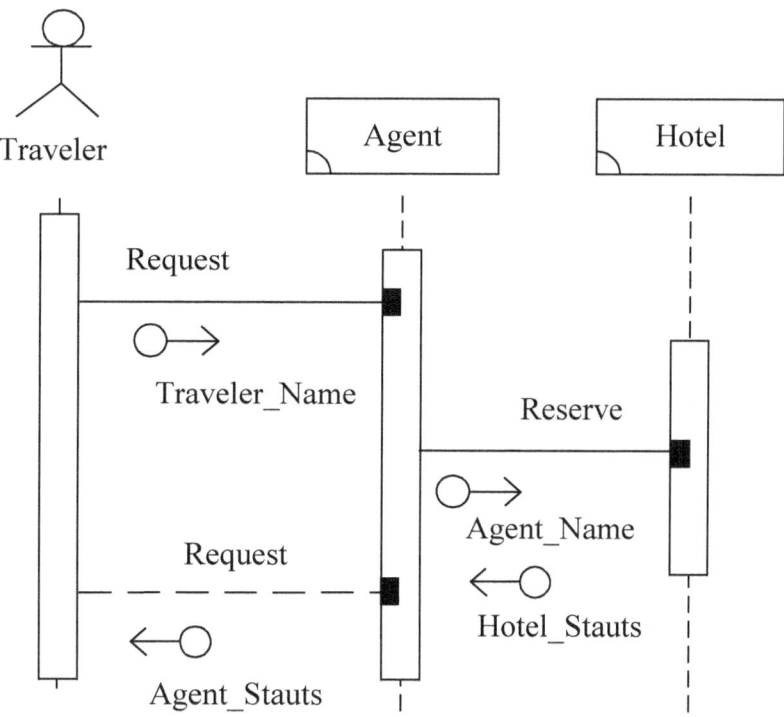

Figure 4-6   Two Interactions Have the Identical Operation

The execution path of Figure 4-6 is as follows. First, external environment's actor *Traveler* interacts with the *Agent* component through the *Request* operation call interaction, carrying the *Traveler_Name* input parameter. Next, component *Agent* interacts with the *Hotel* component through the *Reserve* operation call interaction, carrying the *Agent_Name* input parameter and *Hotel_Stauts* output parameter. Finally, external environment's actor *Traveler* interacts with the *Agent* component through the *Request* operation return interaction, carrying the *Agent_Stauts* output parameter.

An interaction flow diagram may contain a conditional expression. Figure 4-7 shows such an example which has the following execution path. First, external environment's actor *Employee* interacts with the *Computer* component through the *Open* operation call interaction, carrying the *Task_No* input parameter. Next, if the *var_1 < 4 & var_2 > 7* condition is true then component *Computer* shall interact with the *Skype* component through the *Op_1* operation call interaction and component *Skype* shall interact with the *Earphone* component through the *Op_4* operation call interaction, carrying the *Skype_Earphone* output parameter; else if the *var_3 = 99* condition is true then component *Computer* shall interact with the *Skype* component through the *Op_2* operation call interaction and component *Skype* shall interact with the *Speaker* component through the *Op_5* operation call interaction, carrying the *Skype_Speaker* output parameter; else component *Computer* shall interact with the *Youtube* component through the *Op_3* operation call interaction and component *Youtube* shall interact with the *Speaker* component through the *Op_6* operation call interaction, carrying the *Youtube_Speaker* output parameter. Continuingly, if the *var_1 < 4 & var_2 > 7* condition is true then component *Computer* shall interact with the *Skype* component through the *Op_1* operation return interaction, carrying the *Status_1* output parameter; else if the *var_3 = 99* condition is true then component *Computer* shall interact with the *Skype* component through the *Op_2* operation return interaction, carrying the *Status_2* output parameter; else component *Computer* shall interact with the *Youtube* component through the *Op_3* operation return interaction, carrying the *Status_3* output parameter. Finally, external environment's actor *Employee* interacts with the *Computer* component through the *Open* operation return interaction, carrying the *Status* output parameter.

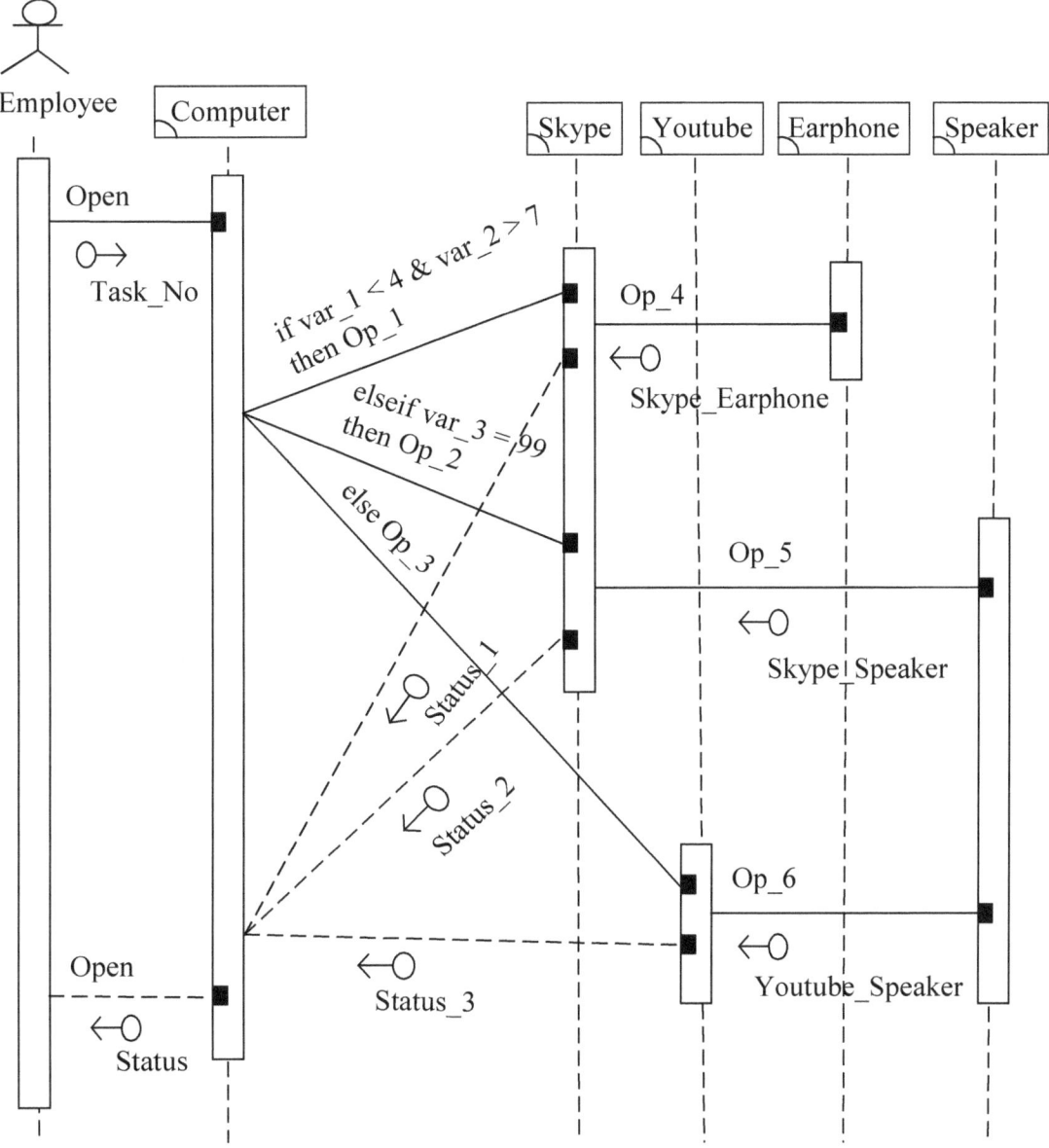

Figure 4-7   Conditional Interaction

Several Boolean conditions are shown in Figure 4-7. They are "*var_1 < 4 &*
var_2 > 7" and "*var_3 = 99*". Variables, such as *var_1*, *var_2* and *var_3*, appearing
in the Boolean condition can be local or global variables [Prat00, Seth96].

PART III: SYSTEMS ARCHITECTURE OF SHSCASIS

Chapter 5: AHD of the SHSCASIS

AHD is the architecture hierarchy diagram we obtain after the architecture construction is finished. Figure 5-1 shows an AHD of the *Smart Home Security Cloud Applications and Services IoT System* (SHSCASIS). In the figure, *SHSCASIS* is composed of *Application_Layer*, *Data_Layer* and *Technology_Layer*; *Application_Layer* is composed of *Presentation_Layer* and *Logic_Layer*; *Presentation_Layer* is composed of *Home_Account_Registering_UI*, *Alerts_Notifying_UI*, *Emergency_Response_Starting_Time_UI* and *Emergency_Response_End_Time_UI*; *Logic_Layer* is composed of *Intrusion_Signs_Daemon*; *Data_Layer* is composed of *SHSCASIS_Database*; *Technology_Layer* is composed of *Intrusion_Signs_Sensor_R (R = AAA000 to ZZZ999)*.

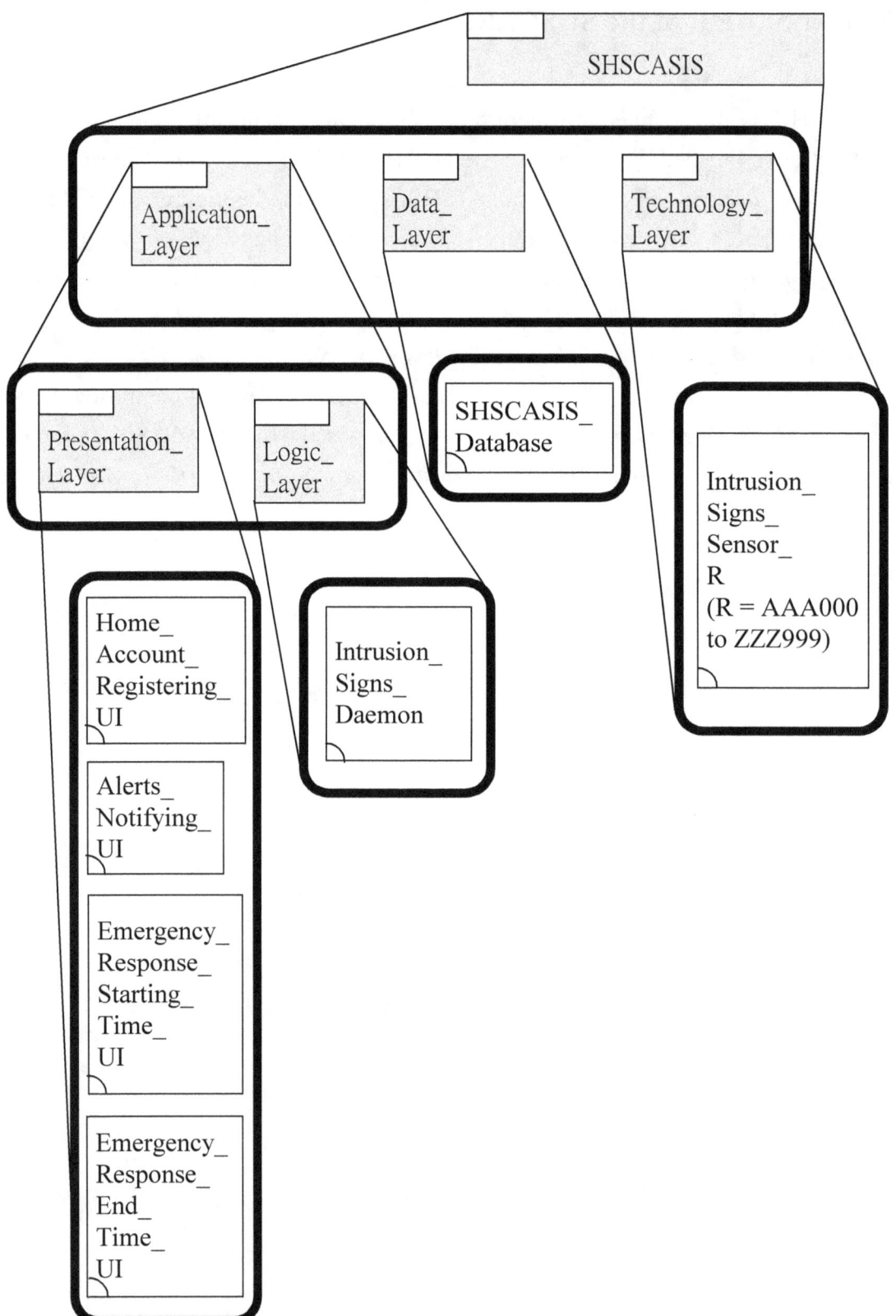

Figure 5-1   AHD of the *SHSCASIS*

In Figure 5-1, *SHSCASIS, Application_Layer, Presentation_Layer, Logic_Layer, Data_Layer* and *Technology_Layer* are aggregated systems while *Home_Account_Registering_UI,* *Alerts_Notifying_UI, Emergency_Response_Starting_Time_UI,* *Emergency_Response_End_Time_UI, Intrusion_Signs_Daemon, SHSCASIS_Database, Intrusion_Signs_Sensor_R (R = AAA000 to ZZZ999)* are non-aggregated systems.

Chapter 6: FD of the SHSCASIS

FD is the framework diagram we obtain after the architecture construction is finished. Figure 6-1 shows a FD of the *Smart Home Security Cloud Applications and Services IoT System* (SHSCASIS).

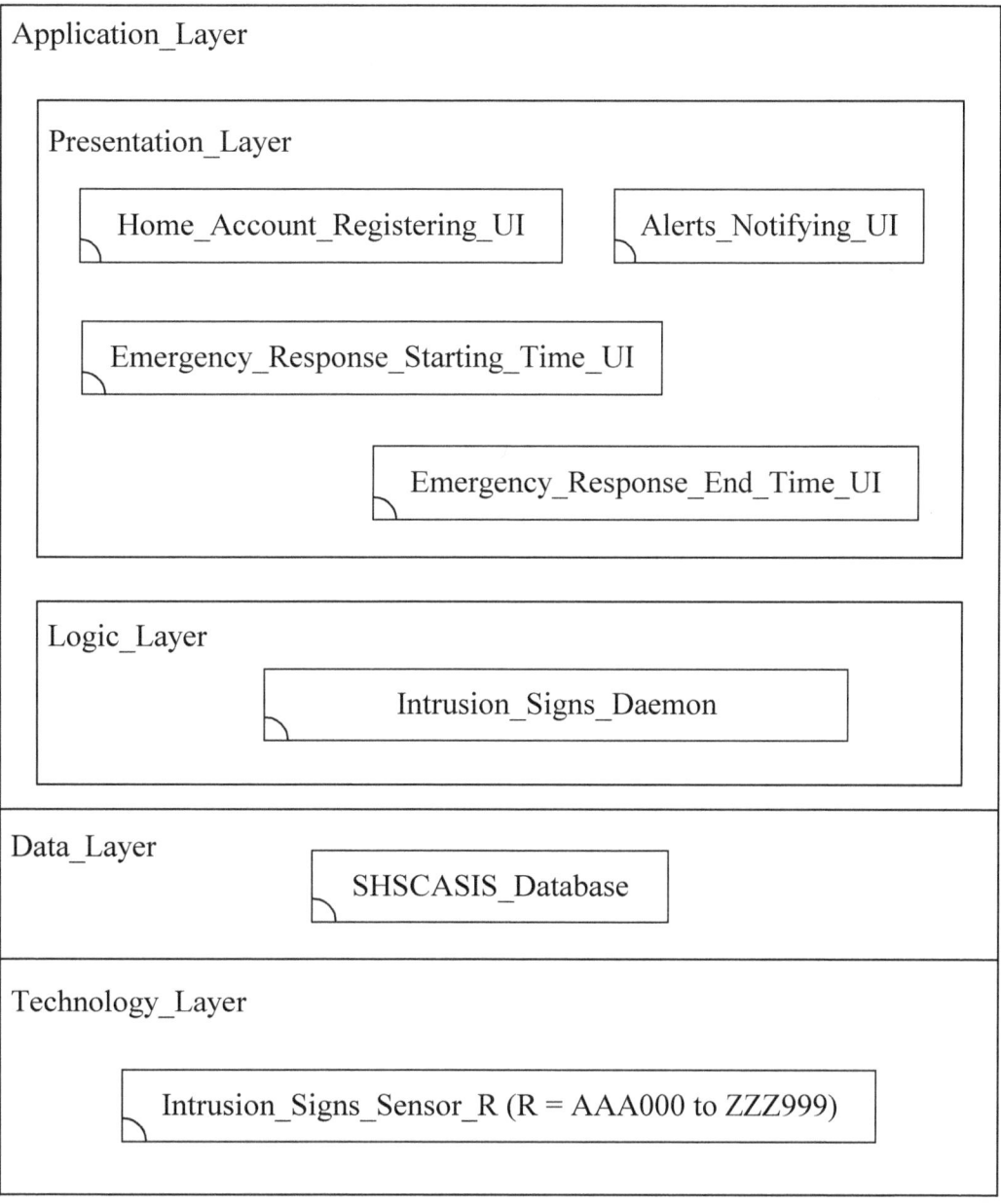

Figure 6-1   FD of the *SHSCASIS*

82

In the above figure, *Presentation_Layer* and *Logic_Layer* are sub-layers of *Application_Layer*. *Presentation_Layer* contains the *Home_Account_Registering_UI*, *Alerts_Notifying_UI*, *Emergency_Response_Starting_Time_UI* and *Emergency_Response_End_Time_UI* components; *Logic_Layer* contains the *Intrusion_Signs_Daemon* component; *Data_Layer* contains the *SHSCASIS_Database* component; *Technology_Layer* contains the *Intrusion_Signs_Sensor_R (R = AAA000 to ZZZ999)* components.

Chapter 7: COD of the SHSCASIS

COD is the component operation diagram we obtain after the architecture construction is finished. Figure 7-1 shows a COD of the *Smart Home Security Cloud Applications and Services IoT System* (SHSCASIS). In the figure, component *Home_Account_Registering_UI* has one operation: *Input_Home_Account*; component *Intrusion_Signs_Daemon* has one operation: *Fork_ISD_Process*; component *Alerts_Notifying_UI* has one operation: *Show_All_Alerts*; component *Emergency_Response_Starting_Time_UI* has one operation: *Input_Emergency_Response_Starting_Time*; component *Emergency_Response_End_Time_UI* has one operation: *Input_Emergency_Response_End_Time*; component *SHSCASIS_Database* has five operations: *SQL_Insert_Home_Account*, *SQL_Insert_Intrusion_Signs*, *SQL_Select_Intrusion_Signs_for_Alerts_Analysis*, *SQL_Insert_Emergency_Response_Starting_Time* and *SQL_Insert_Emergency_Response_End_Time*; component *Intrusion_Signs_Sensor_R (R = AAA000 to ZZZ999)* has two operations: *Sense_Intrusion_Signs* and *Return_Intrusion_Signs*.

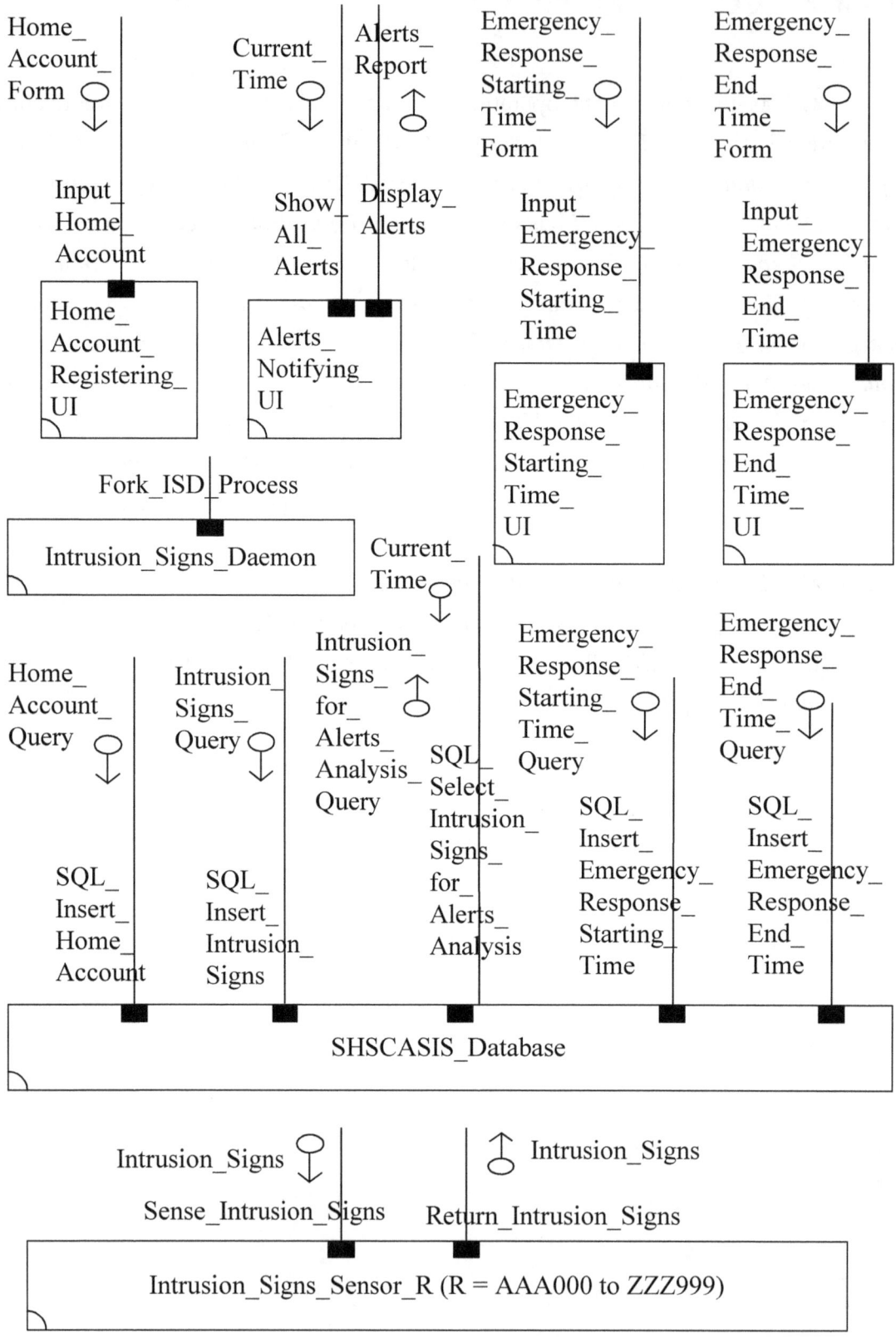

Figure 7-1   COD of the *SHSCASIS*

The operation formula of *Input_Home_Account* is *Input_Home_Account(In Home_Account_Form)*. The operation formula of *Fork_ISD_Process* is *Fork_ISD_Process*. The operation formula of *Show_All_Alerts* is *Show_All_Alerts(In Current_Time)*. The operation formula of *Display_Alerts* is *Display_Alerts(Out Alerts_Report)*. The operation formula of *Input_Emergency_Response_Starting_Time* is *Input_Emergency_Response_Starting_Time(In Emergency_Response_Starting_Time_Form)*. The operation formula of *Input_Emergency_Response_End_Time* is *Input_Emergency_Response_End_Time(In Emergency_Response_End_Time_Form)*. The operation formula of *SQL_Insert_Home_Account* is *SQL_Insert_Home_Account(In Home_Account_Query)*. The operation formula of *SQL_Insert_Intrusion_Signs* is *SQL_Insert_Intrusion_Signs(In Intrusion_Signs_Query)*. The operation formula of *SQL_Select_Intrusion_Signs_for_Alerts_Analysis* is *SQL_Select_Intrusion_Signs_for_Alerts_Analysis(In Current_Time; Out Intrusion_Signs_for_Alerts_Analysis_Query)*. The operation formula of *SQL_Insert_Emergency_Response_Starting_Time* is *SQL_Insert_Emergency_Response_Starting_Time(In Emergency_Response_Starting_Time_Query)*. The operation formula of *SQL_Insert_Emergency_Response_End_Time* is *SQL_Insert_Emergency_Response_End_Time(In Emergency_Response_End_Time_Query)*. The operation formula of *Sense_Intrusion_Signs* is *Sense_Intrusion_Signs(In Intrusion_Signs)*. The operation formula of *Return_Intrusion_Signs* is *Return_Intrusion_Signs(Out Intrusion_Signs)*.

Figure 7-2 shows the composite data type specification of the *Home_Account_Form* input parameter occurring in the *Input_Home_Account(In Home_Account_Form)* operation formula.

	Home_Account_Form
Data Type	TABLE of Home_Number: Text Address: Text Owner_Name: Text Owner_Phone: Text End TABLE ;
Instances	SHSCASIS *Smart Home Security* **Home Account Form** Home_Number: B205144699 Address: 122 Antonia Drive, Niskayuna, NY 12309 Owner_Name: Deborah Henson Owner_Phone: 517-402-7733

Figure 7-2 Composite Data Type Specification of *Home_Account_Form*

Figure 7-3 shows the primitive data type specification of the *Current_Time* parameter occurring in the *Show_All_Alerts(In Current_Time)* and *SQL_Select_Intrusion_Signs_for_Alerts_Analysis(In Current_Time; Out Intrusion_Signs_for_Alerts_Analysis_Query)* operation formulas.

Parameter	Data Type	Instances
Current_ Time	Text	2016/07/03, 15:30 PM

Figure 7-3   Primitive Data Type Specification

88

Figure 7-4 shows the composite data type specification of the *Alerts_Report* output parameter occurring in the *Display_Alerts(Out Alerts_Report)* operation formula.

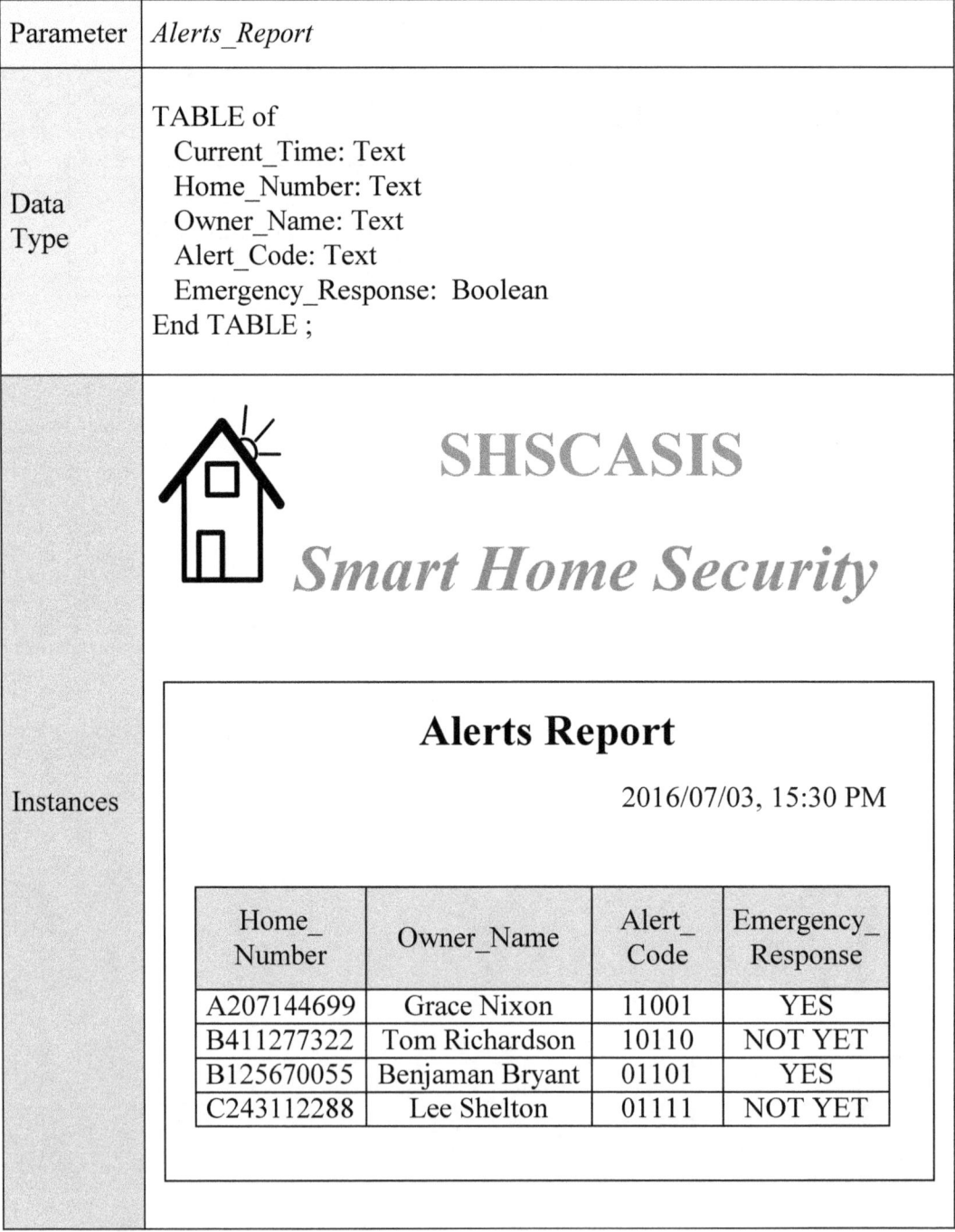

Parameter	*Alerts_Report*

Data Type	TABLE of
	    Current_Time: Text
	    Home_Number: Text
	    Owner_Name: Text
	    Alert_Code: Text
	    Emergency_Response: Boolean
	End TABLE ;

Figure 7-4   Composite Data Type Specification of *Alerts_Report*

Figure 7-5 shows the composite data type specification of the *Emergency_Response_Starting_Time_Form* input parameter occurring in the *Input_Emergency_Response_Starting_Time(In Emergency_Response_Starting_Time_Form)* operation formula.

Parameter	*Emergency_Response_Starting_Time_Form*

Data Type	TABLE of
	  Home_Number: Text
	  Owner_Name: Text
	  Emergency_Response_Starting_Time: Text
	  Actions: Text
	End TABLE ;

Instances	

Figure 7-5   Composite Data Type Specification of
Emergency_Response_Starting_Time_Form

Figure 7-6 shows the composite data type specification of the *Emergency_Response_End_Time_Form* input parameter occurring in the *Input_Emergency_Response_End_Time(In Emergency_Response_End_Time_Form)* operation formula.

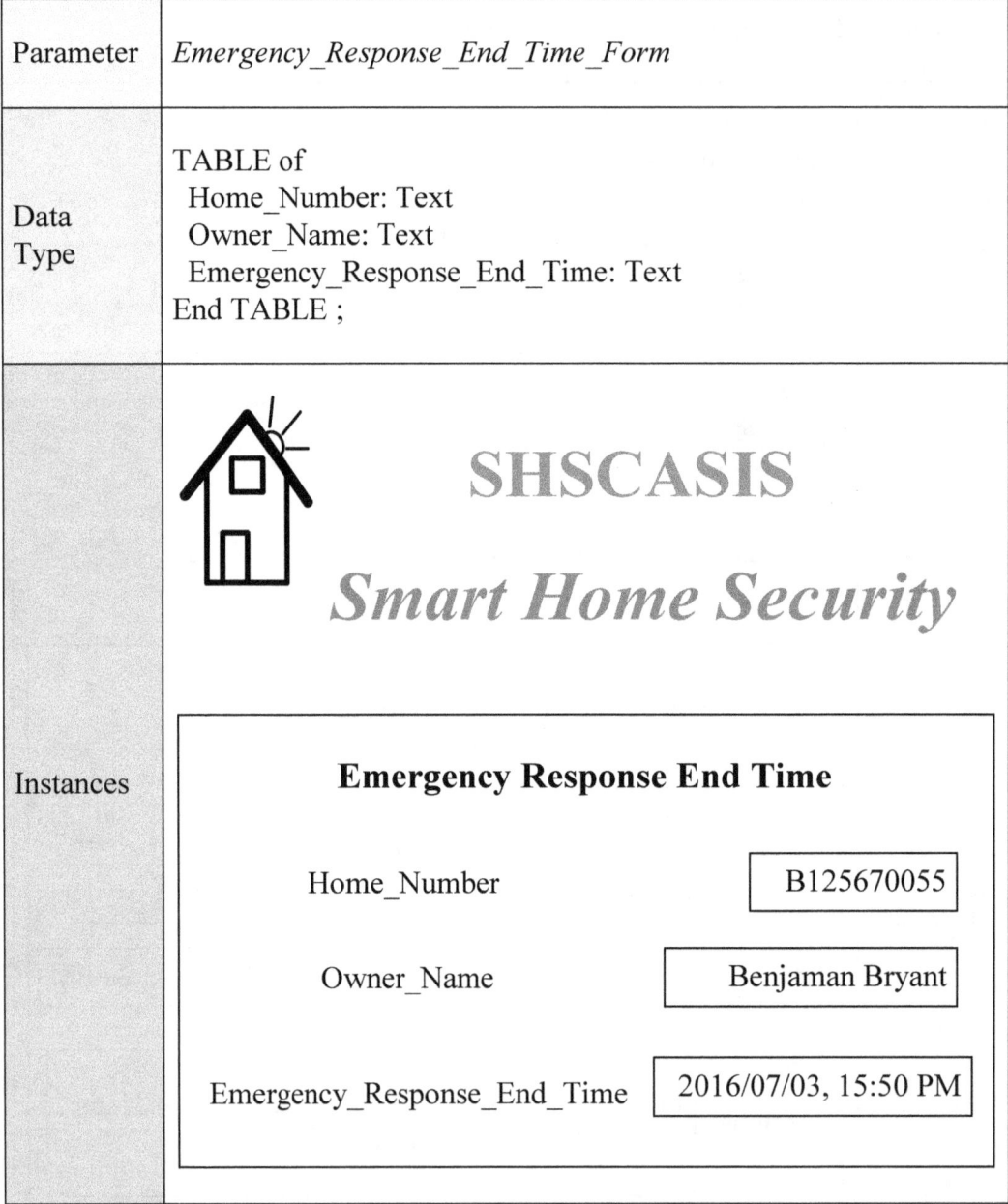

Parameter	*Emergency_Response_End_Time_Form*

Data Type	TABLE of
	   Home_Number: Text
	   Owner_Name: Text
	   Emergency_Response_End_Time: Text
	 End TABLE ;

Figure 7-6   Composite Data Type Specification of
Emergency_Response_End_Time_Form

Figure 7-7 shows the composite data type specification of the *Home_Account_Query* input parameter occurring in the *SQL_Insert_Home_Account(In Home_Account_Query)* operation formula.

Parameter	*Home_Account_Query*

Data Type	TABLE of
	  Home_Number: Text
	  Address: Text
	  Owner_Name: Text
	  Owner_Phone: Text
	End TABLE ;

Instances	<table><tr><td>Home_ Number</td><td>Address</td></tr><tr><td>B205144699</td><td>122 Antonia Drive, Niskayuna, NY 12309</td></tr></table> <table><tr><td>Owner_ Name</td><td>Owner_ Phone</td></tr><tr><td>Deborah Henson</td><td>517-402-7733</td></tr></table>

Figure 7-7   Composite Data Type Specification of
Home_Account_Query

Figure 7-8 shows the composite data type specification of the *Intrusion_Signs_Query* input parameter occurring in the *SQL_Insert_Intrusion_Signs(In Intrusion_Signs_Query)* operation formula.

Parameter	*Intrusion_Signs_Query*

Data Type	TABLE of
	  Home_Number: Text
	  Sensing_Time: Text
	  Doors_Open: Boolean
	  Windows_Open: Boolean
	  Motions: Boolean
	  Sound: Boolean
	  Vibration: Boolean
	End TABLE ;

Instances	(see below)

Home_Number	Sensing_Time
B411277322	2016/07/03, 15:28 PM

Doors_Open	Windows_Open	Motions	Sound	Vibration
YES	NO	YES	YES	NO

Figure 7-8   Composite Data Type Specification
of *Intrusion_Signs_Query*

Figure 7-9 shows the composite data type specification of the *Intrusion_Signs_for_Alerts_Analysis_Query* output parameter occurring in the *SQL_Select_Intrusion_Signs_for_Alerts_Analysis(In Current_Time; Out Intrusion_Signs_for_Alerts_Analysis_Query)* operation formula.

Parameter	*Intrusion_Signs_for_Alerts_Analysis_Query*

Data Type	TABLE of
	  Home_Number: Text
	  Doors_Open: Boolean
	  Windows_Open: Boolean
	  Motions: Boolean
	  Sound: Boolean
	  Vibration: Boolean
	  Emergency_Response_Starting_Time: Text
	End TABLE ;

Instances	(see tables below)

Home_ Number	Doors_ Open	Windows_ Open	Motions
A207144699	YES	YES	NO
B411277322	YES	NO	YES
B125670055	NO	YES	YES
C243112288	NO	YES	YES

Sound	Vibration	Emergency_ Response_ Starting_ Time
NO	YES	2016/07/03, 15:22
YES	NO	NULL
NO	YES	2016/07/03, 15:24
YES	YES	NULL

Figure 7-9   Composite Data Type Specification of
Intrusion_Signs_for_Alerts_Analysis_Query

Figure 7-10 shows the composite data type specification of the *Emergency_Response_Starting_Time_Query* input parameter occurring in the *SQL_Insert_Emergency_Response_Starting_Time(In Emergency_Response_Starting_Time_Query)* operation formula.

Parameter	*Emergency_Response_Starting_Time_Query*

Data Type	TABLE of
	  Home_Number: Text
	  Owner_Name: Text
	  Emergency_Response_Starting_Time: Text
	  Actions: Text
	End TABLE ;

Instances	<table><tr><td>Home_ Number</td><td>Owner_ Name</td><td>Emergency_ Response_ Starting_ Time</td></tr><tr><td>B125670055</td><td>Benjaman Bryant</td><td>2016/07/03, 15:24</td></tr></table> <table><tr><td>Actions</td></tr><tr><td>Inform the house owner about the intrusion signs; Contact law enforcement.</td></tr></table>

Figure 7-10   Composite Data Type Specification
of *Emergency_Response_Starting_Time_Query*

Figure 7-11 shows the composite data type specification of the *Emergency_Response_End_Time_Query* input parameter occurring in the *SQL_Insert_Emergency_Response_End_Time(In Emergency_Response_End_Time_Query)* operation formula.

Parameter	*Emergency_Response_End_Time_Query*
Data Type	TABLE of Home_Number: Text Owner_Name: Text Emergency_Response_End_Time: Text End TABLE ;
Instances	

Home_ Number	Owner_ Name:	Emergency_ Response_ End_ Time
B125670055	Benjaman Bryant	2016/07/03, 15:50

Figure 7-11   Composite Data Type Specification
of *Emergency_Response_End_Time_Query*

Figure 7-12 shows the primitive data type specification of the *Intrusion_Signs* parameter occurring in the *Sense_Intrusion_Signs(In Intrusion_Signs)* and *Return_Intrusion_Signs(Out Intrusion_Signs)* operation formulas.

Parameter	*Intrusion_Signs*				

Data Type	TABLE of
	  Doors_Open: Boolean
	  Windows_Open: Boolean
	  Motions: Boolean
	  Sound: Boolean
	  Vibration: Boolean
	End TABLE ;				

Instances	Doors_Open	Windows_Open	Motions	Sound	Vibration
	YES	NO	YES	YES	NO

Figure 7-12   Composite Data Type Specification of
Intrusion_Signs

Chapter 8: CCD of the SHSCASIS

CCD is the component connection diagram we obtain after the architecture construction is finished. Figure 8-1 shows a CCD of the *Smart Home Security Cloud Applications and Services IoT System* (SHSCASIS).

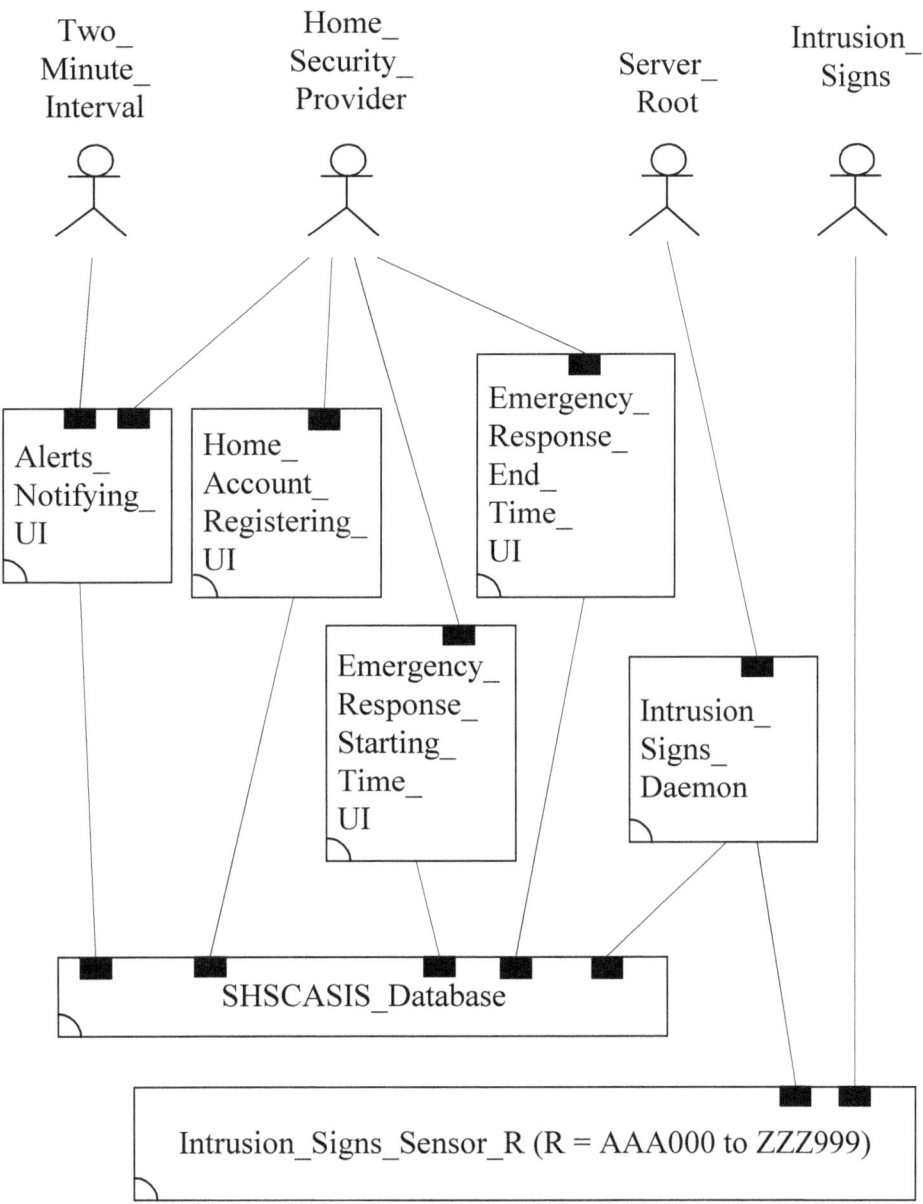

Figure 8-1   CCD of the *SHSCASIS*

In the above figure, actor *Two_Minute_Interval* has a connection with the *Alerts_Notifying_UI* component; actor *Home_Security_Provider* has a connection with each one of the *Alerts_Notifying_UI, Home_Account_Registering_UI, Emergency_Response_Starting_Time_UI, Emergency_Response_End_Time_UI* components; actor *Server_Root* has a connection with the *Intrusion_Signs_Daemon* components; actor *Intrusion_Signs* has a connection with the *Intrusion_Signs_Sensor_R (R = AAA000 to ZZZ999)* component; each one of the *Alerts_Notifying_UI, Home_Account_Registering_UI, Emergency_Response_Starting_Time_UI, Emergency_Response_End_Time_UI* components has a connection with the *SHSCASIS_Database* component; component *Intrusion_Signs_Daemon* has a connection with each one of the *SHSCASIS_Database, Intrusion_Signs_Sensor_R (R = AAA000 to ZZZ999)* components.

Chapter 9: SBCD of the SHSCASIS

SBCD is the structure-behavior coalescence diagram we obtain after the architecture construction is finished. Figure 9-1 shows a SBCD of the *Smart Home Security Cloud Applications and Services IoT System* (SHSCASIS) in which interactions among the *Two_Minute_Interval, Home_Security_Provider, Server_Root, Intrusion_Signs* actors and the *Alerts_Notifying_UI, Home_Account_Registering_UI, Emergency_Response_Starting_Time_UI,* *Emergency_Response_End_Time_UI, Intrusion_Signs_Daemon, SHSCASIS_Database, Intrusion_Signs_Sensor_R (R = AAA000 to ZZZ999)* components shall draw forth the *Registering_Home_Account, Sensing_Intrusion_Signs,* *Alerts_Notifying, Recording_Emergency_Response_Starting_Time, Recording_Emergency_Response_End_Time* behaviors.

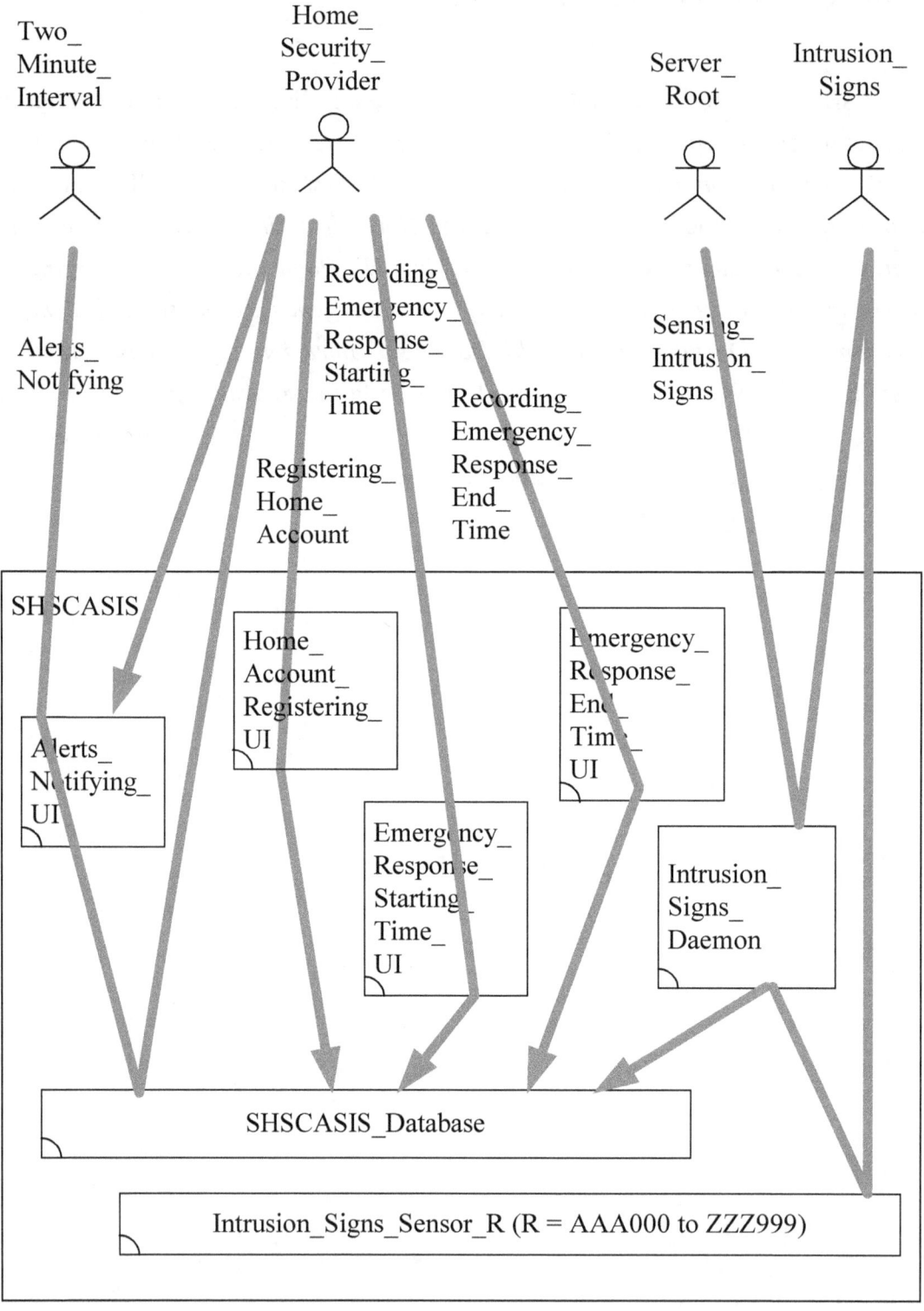

Figure 9-1   SBCD of the *SHSCASIS*

The overall behavior of the *Smart Home Security Cloud Applications and Services IoT System* includes the *Registering_Home_Account*, *Sensing_Intrusion_Signs*, *Alerts_Notifying*, *Recording_Emergency_Response_Starting_Time*, *Recording_Emergency_Response_End_Time* behaviors. In other words, the *Registering_Home_Account*, *Sensing_Intrusion_Signs*, *Alerts_Notifying*, *Recording_Emergency_Response_Starting_Time*, *Recording_Emergency_Response_End_Time* behaviors together provide the overall behavior of the *Smart Home Security Cloud Applications and Services IoT System*.

The major purpose of adopting the architectural approach, instead of separating the structure model from the behavior model, is to achieve one single coalesced model. In Figure 9-1, systems architects are able to see that the systems structure and systems behavior coexist in the SBCD. That is, in the SBCD of *Smart Home Security Cloud Applications and Services IoT System*, systems architects not only see its systems structure but also see (at the same time) its systems behavior.

Chapter 10: IFD of the SHSCASIS

IFDs are the interaction flow diagrams we obtain after the architecture construction is finished. The overall behavior of the *Smart Home Security Cloud Applications and Services IoT System* (SHSCASIS) includes five individual behaviors: *Registering_Home_Account*, *Sensing_Intrusion_Signs*, *Alerts_Notifying*, *Recording_Emergency_Response_Starting_Time* and *Recording_Emergency_Response_End_Time*. Each individual behavior is represented by an execution path. We use an IFD to define each one of these execution paths.

Figure 10-1 shows an IFD of the *Registering_Home_Account* behavior. First, actor *Home_Security_Provider* interacts with the *Home_Account_Registering_UI* component through the *Input_Home_Account* operation call interaction, carrying the *Home_Account_Form* input parameter. Finally, component *Home_Account_Registering_UI* interacts with the *SHSCASIS_Database* component through the *SQL_Insert_Home_Account* operation call interaction, carrying the *Home_Account_Query* input parameter.

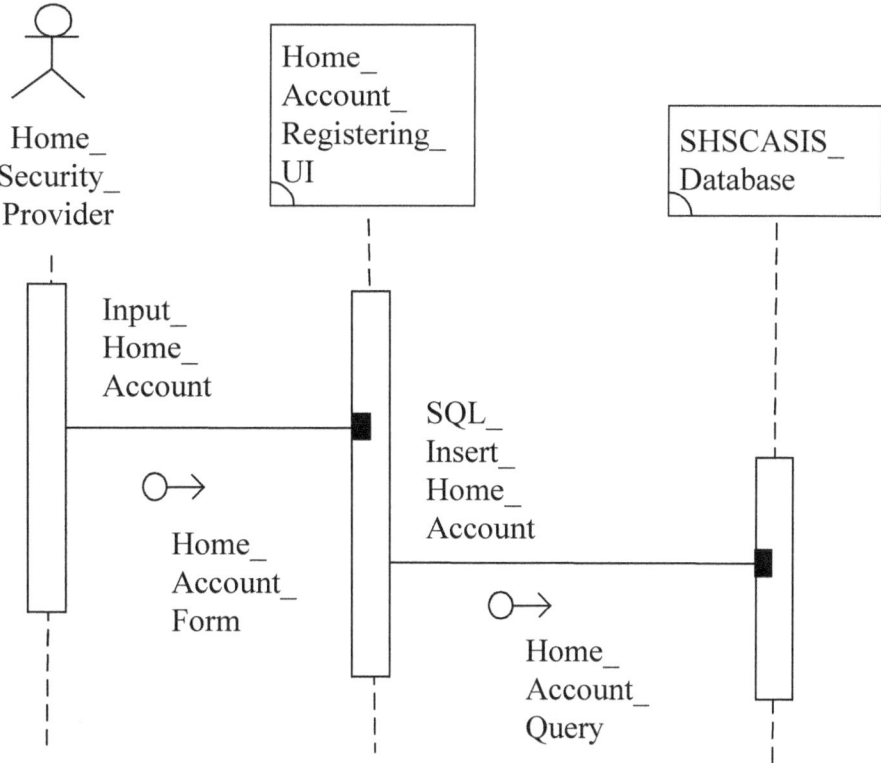

Figure 10-1   IFD of the *Registering_Home_Account* Behavior

Figure 10-2 shows an IFD of the *Sensing_Intrusion_Signs* behavior. First, actor *Server_Root* interacts with the *Intrusion_Signs_Daemon* component through the *Fork_ISD_Process* operation call interaction. Next, actor *Intrusion_Signs* interacts with the *Intrusion_Signs_Sensor_R (R = AAA000 to ZZZ999)* component through the *Sense_Intrusion_Signs* operation call interaction, carrying the *Intrusion_Signs* input parameter. Continuingly, component *Intrusion_Signs_Daemon* interacts with the *Intrusion_Signs_Sensor_R (R = AAA000 to ZZZ999)* component through the *Return_Intrusion_Signs* operation call interaction, carrying the *Intrusion_Signs* output parameter. Finally, component *Intrusion_Signs_Daemon* interacts with the *SHSCASIS_Database* component through the *SQL_Insert_Intrusion_Signs* operation call interaction, carrying the *Intrusion_Signs_Query* input parameter.

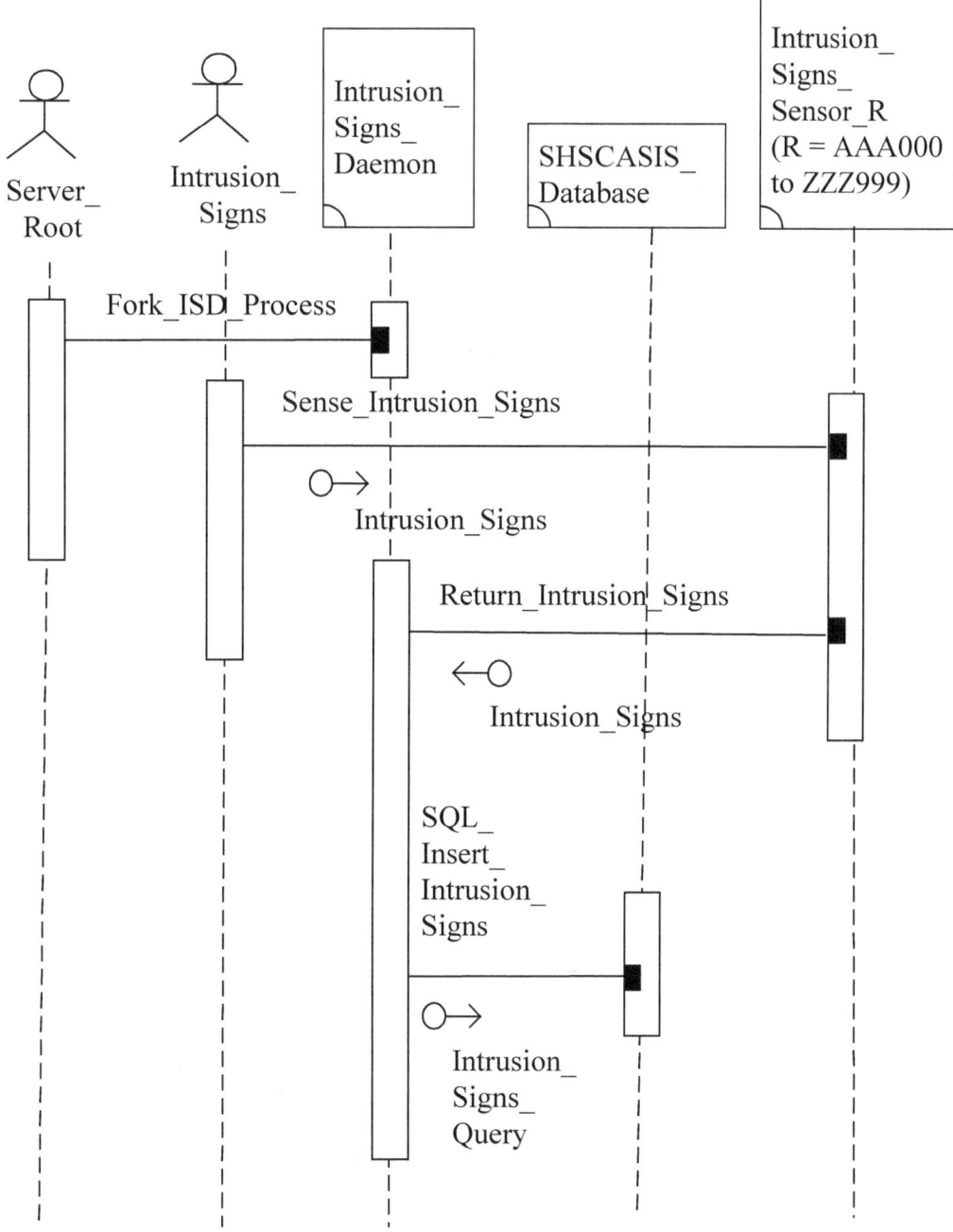

Figure 10-2   IFD of the *Sensing_Intrusion_Signs* Behavior

Figure 10-3 shows an IFD of the *Alerts_Notifying* behavior. First, actor *Two_Minute_Interval* interacts with the *Alerts_Notifying_UI* component through the *Show_All_Alerts* operation call interaction, carrying the *Current_Time* input parameter. Next, component *Alerts_Notifying_UI* interacts with the *SHSCASIS_Database* component through the *SQL_Select_Intrusion_Signs_for_Alerts_Analysis* operation call interaction, carrying the *Current_Time* input parameter and *Intrusion_Signs_for_Alerts_Analysis_Query* output parameter. Finally, actor *Home_Security_Provider* interacts with the *Alerts_Notifying_UI* component through the *Display_Alerts* operation call interaction, carrying the *Alerts_Report* output parameter.

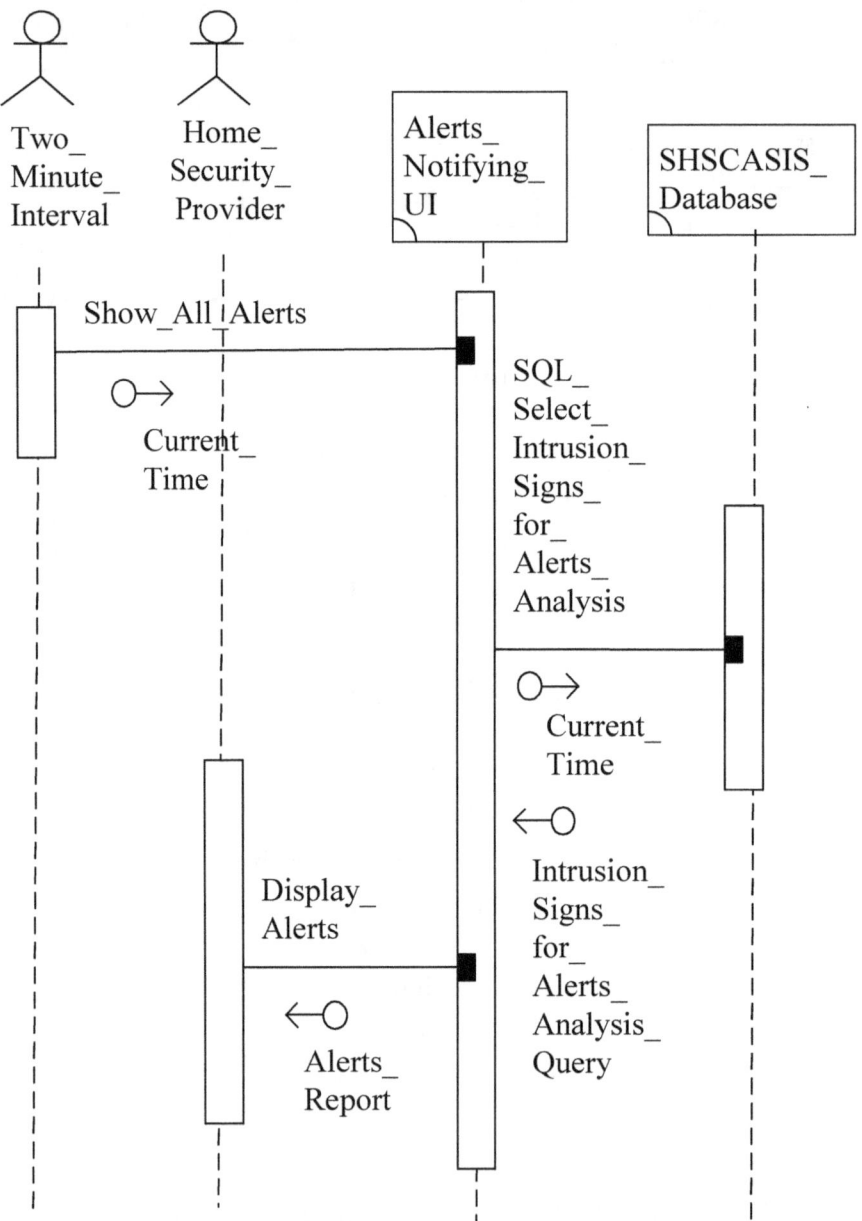

Figure 10-3   IFD of the *Alerts_Notifying* Behavior

Figure 10-4 shows an IFD of the *Recording_Emergency_Response_Starting_Time* behavior. First, actor *Home_Security_Provider* interacts with the *Emergency_Response_Starting_Time_UI* component through the *Input_Emergency_Response_Starting_Time* operation call interaction, carrying the *Emergency_Response_Starting_Time_Form* input parameter. Finally, component *Emergency_Response_Starting_Time_UI* interacts with the *SHSCASIS_Database* component through the *SQL_Insert_Emergency_Response_Starting_Time* operation call interaction, carrying the *Emergency_Response_Starting_Time_Query* input parameter.

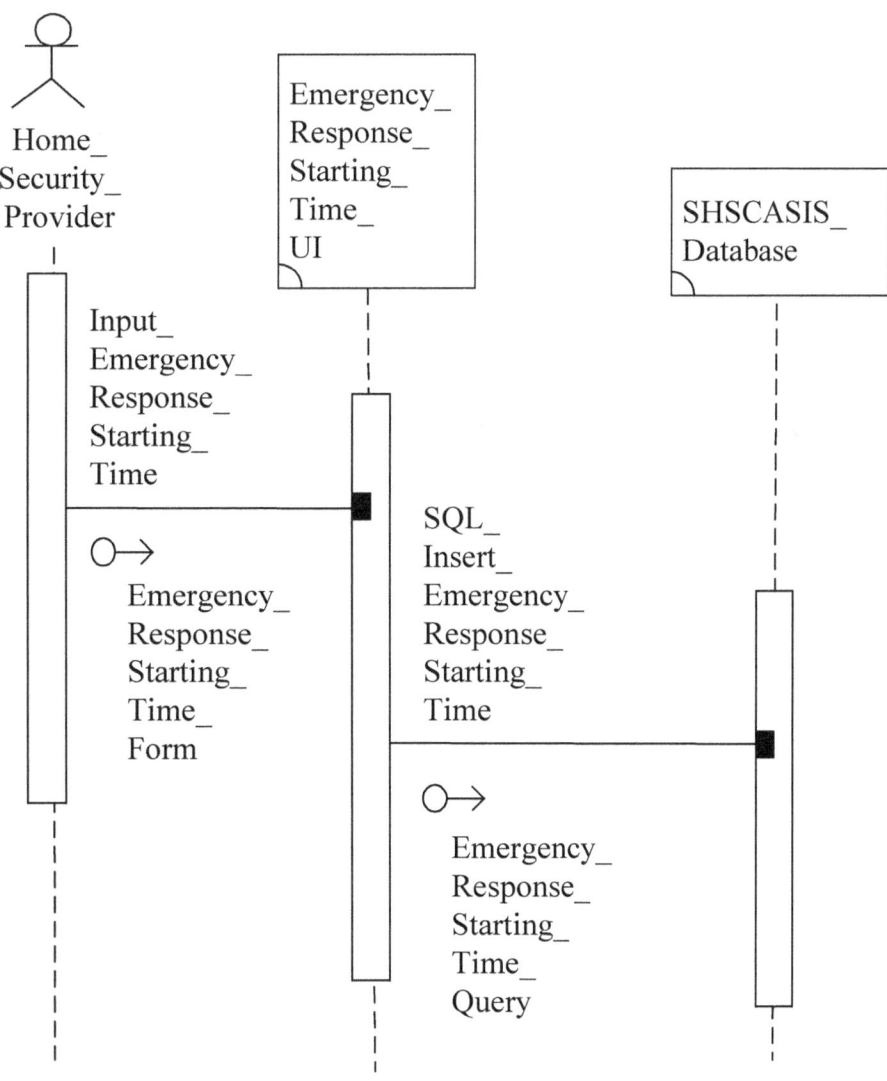

Figure 10-4   IFD of the
Recording_Emergency_Response_Starting_Time Behavior

Figure 10-5 shows an IFD of the *Recording_Emergency_Response_End_Time* behavior. First, actor *Home_Security_Provider* interacts with the *Emergency_Response_End_Time_UI* component through the *Input_Emergency_Response_End_Time* operation call interaction, carrying the *Emergency_Response_End_Time_Form* input parameter. Finally, component *Emergency_Response_End_Time_UI* interacts with the *SHSCASIS_Database* component through the *SQL_Insert_Emergency_Response_End_Time* operation call interaction, carrying the *Emergency_Response_End_Time_Query* input parameter.

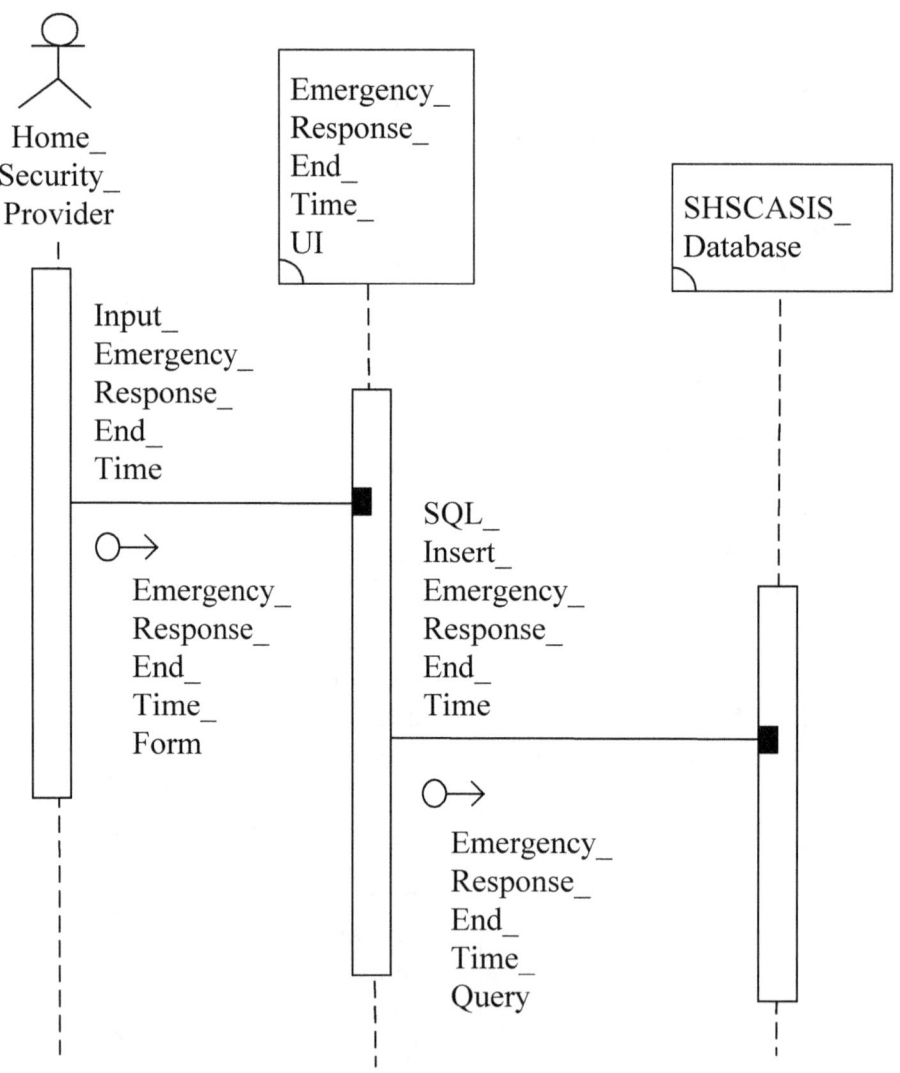

Figure 10-5   IFD of the
Recording_Emergency_Response_End_Time Behavior

APPENDIX A: SBC ARCHITECTURE DESCRIPTION LANGUAGE

(1) Architecture Hierarchy Diagram

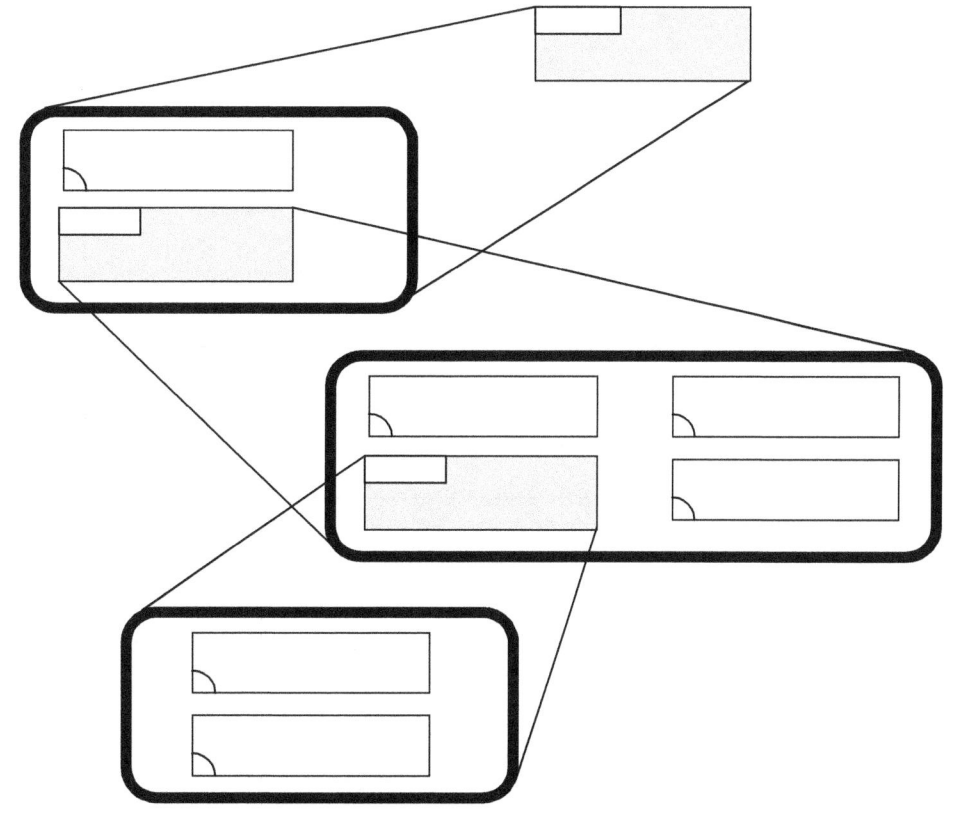

: Aggregated System

: Non-Aggregated System, Component

(2) Framework Diagram

: Component

(3) Component Operation Diagram

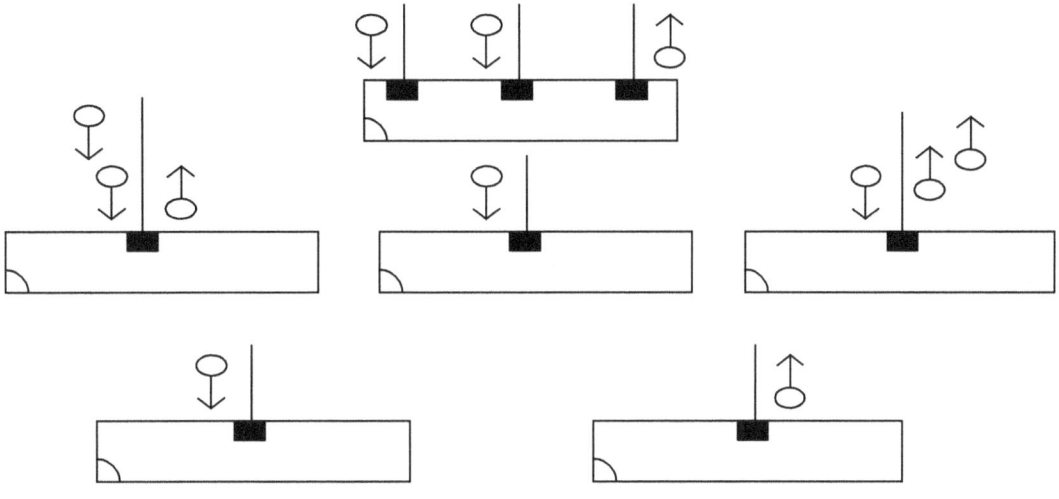

 : Operation

: Input Data

: Output Data

: Component

(4) Component Connection Diagram

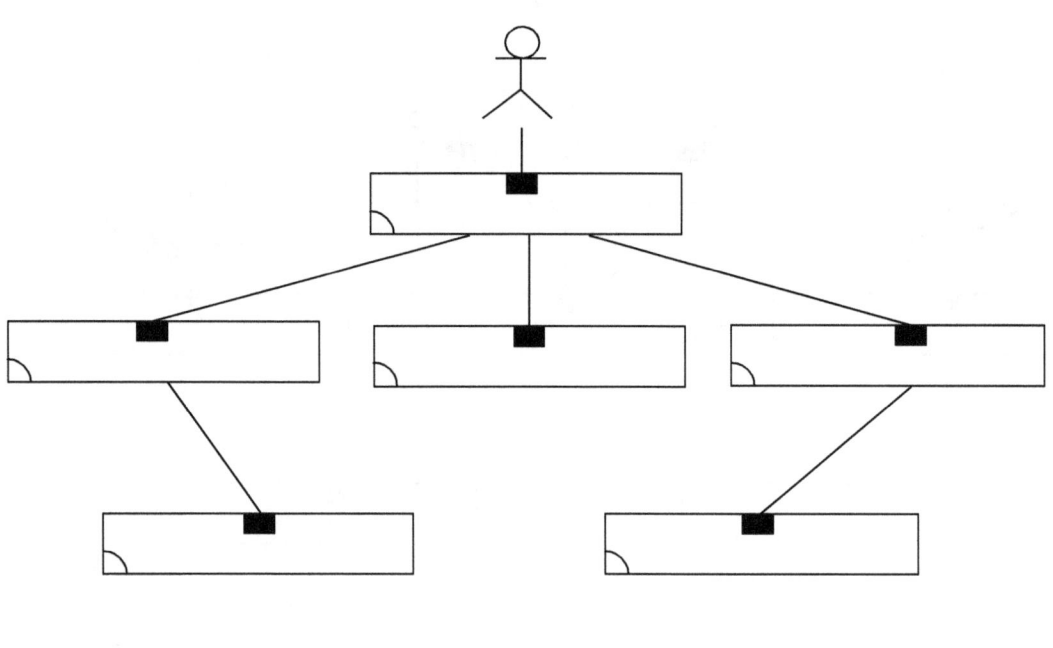

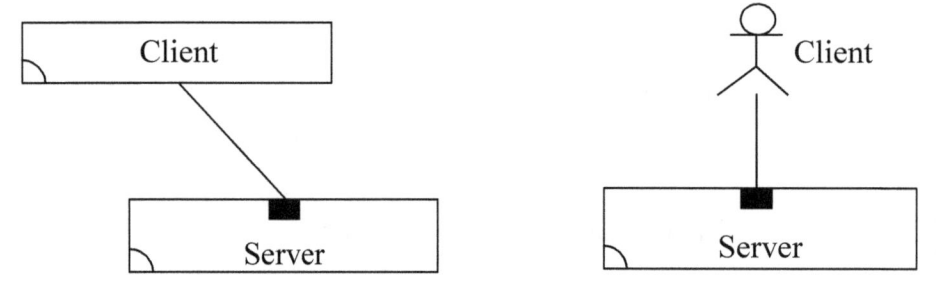

(5) Structure-Behavior Coalescence Diagram

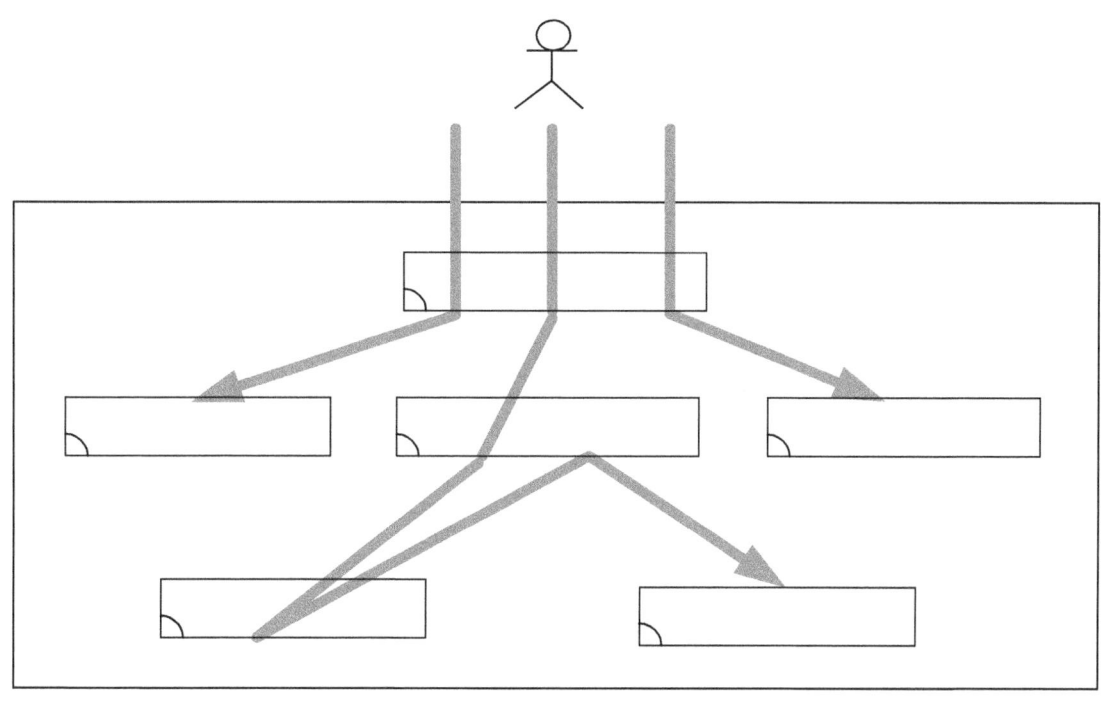

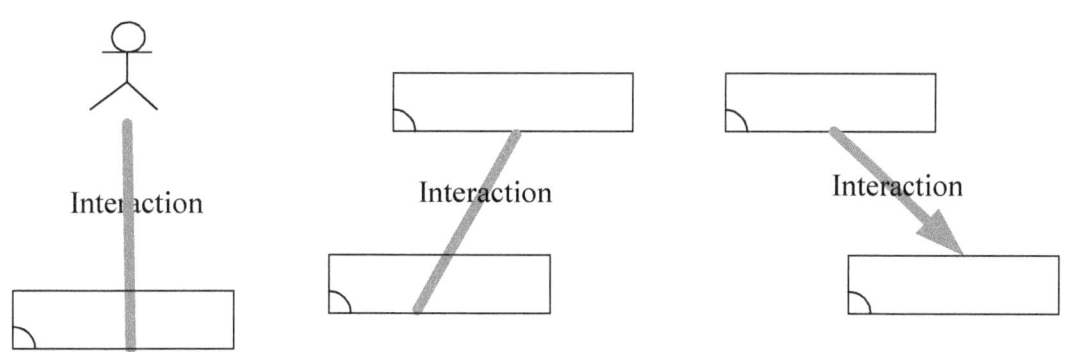

(6) Interaction Flow Diagram

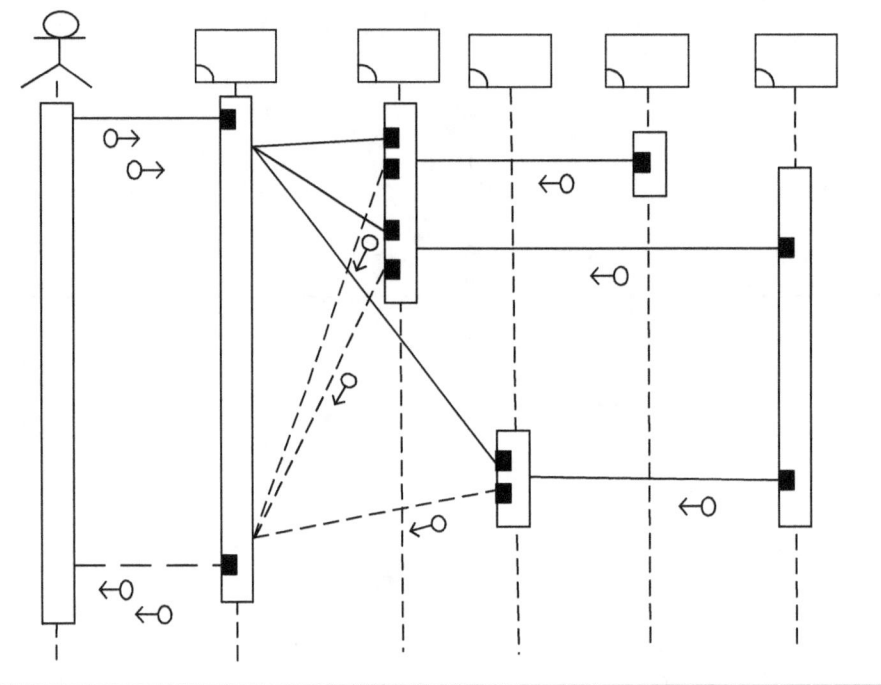

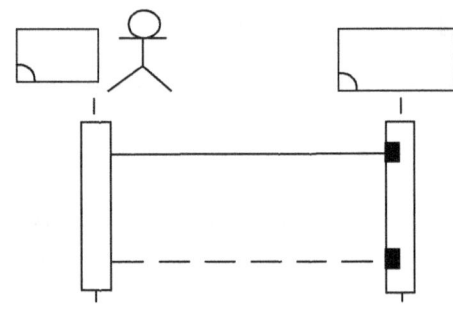

: Operation Call Interaction

: Operation Return Interaction

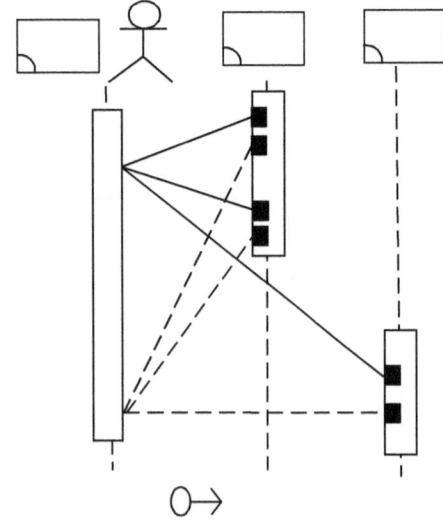

: Conditional
Operation Call Interaction

: Conditional
Operation Return Interaction

○→   : Input Data

←○   : Output Data

APPENDIX B: SBC PROCESS ALGEBRA

(1) Operation-Based Single-Queue SBC Process Algebra

(1) \<System\> ::= **fix**(" \<Process_Variable\> "="\<IFD\> " • " \<Process_Variable\>
{"+" \<IFD\> " • " \<Process_Variable\>} ")"

(2) \<IFD\> ::= \<Type_1_Interaction\> {"• " \<Type_1_Or_2_Interaction\>}

(3) \<Type_1_Or_2_Interaction\> ::= \<Type_1_Interaction\>

| \<Type_2_Interaction\>

(2) Operation-Based Multi-Queue SBC Process Algebra

(1) <System> ::= <FixIFD> {"||" <FixIFD>}

(2) <FixIFD> ::= "**fix**(" <Process_Variable>"="<IFD>
           "●" <Process_Variable> ")"

(3) <IFD> ::= <Type_1_Interaction> {"● " Type_1_Or_2_Interaction>}

(4) <Type_1_Or_2_Interaction> ::= <Type_1_Interaction>

                         | <Type_2_Interaction>

(3) Operation-Based Infinite-Queue SBC Process Algebra

(1) <System>   ::=   "! ("<IFD> " • " *STOP* ")" {"‖ ! (" <IFD> " • " *STOP* ")"}

(2) <IFD> ::= <Type_1_Interaction> {"• " <Type_1_Or_2_Interaction>}

(3) <Type_1_Or_2_Interaction> ::= <Type_1_Interaction>

         | <Type_2_Interaction>

BIBLIOGRAPHY

[Bern09] Bernstein, D. et al., "Blueprint for the Intercloud – Protocols and Formats for Cloud Computing Interoperability," *IEEE Computer Society*, 2009, pp. 328-336.

[Burd10] Burd, S. D., *Systems Architecture*, 6th Edition, Cengage Learning, 2010.

[Chao14a] Chao, W. S., *Systems Thingking 2.0: Architectural Thinking Using the SBC Architecture Description Language*, CreateSpace Independent Publishing Platform, 2014.

[Chao14b] Chao, W. S., *General Systems Theory 2.0: General Architectural Theory Using the SBC Architecture*, CreateSpace Independent Publishing Platform, 2014.

[Chao14c] Chao, W. S., *Systems Modeling and Architecting: Structure-Behavior Coalescence for Systems Architecture*, CreateSpace Independent Publishing Platform, 2014.

[Chao15a] Chao, W. S., *A Process Algebra For Systems Architecture: The Structure-Behavior Coalescence Approach*, CreateSpace Independent Publishing Platform, 2015.

[Chao15b] Chao, W. S., *An Observation Congruence Model For Systems Architecture: The Structure-Behavior Coalescence Approach*, CreateSpace Independent Publishing Platform, 2015.

[Chao16] Chao, W. S., *System: Contemporary Concept, Definition, and Language*, CreateSpace Independent Publishing Platform, 2016.

[Chao17a] Chao, W. S., *Channel-Based Single-Queue SBC Process Algebra For Systems Definition: General Architectural Theory at Work*, CreateSpace

Independent Publishing Platform, 2017.

[Chao17b] Chao, W. S., *Channel-Based Multi-Queue SBC Process Algebra For Systems Definition: General Architectural Theory at Work*, CreateSpace Independent Publishing Platform, 2017.

[Chao17c] Chao, W. S., *Channel-Based Infinite-Queue SBC Process Algebra For Systems Definition: General Architectural Theory at Work*, CreateSpace Independent Publishing Platform, 2017.

[Chao17d] Chao, W. S., *Operation-Based Single-Queue SBC Process Algebra For Systems Definition: General Architectural Theory at Work*, CreateSpace Independent Publishing Platform, 2017.

[Chao17e] Chao, W. S., *Operation-Based Multi-Queue SBC Process Algebra For Systems Definition: Unification of Systems Structure and Systems Behavior*, CreateSpace Independent Publishing Platform, 2017.

[Chao17f] Chao, W. S., *Operation-Based Infinite-Queue SBC Process Algebra For Systems Definition: Unification of Systems Structure and Systems Behavior*, CreateSpace Independent Publishing Platform, 2017.

[Chec99] Checkland, P., *Systems Thinking, Systems Practice: Includes a 30-Year Retrospective*, 1st Edition, Wiley, 1999.

[Craw15] Crawley, P. et al., *System Architecture: Strategy and Product Development for Complex Systems*, Prentice Hall, 2015.

[Dam06] Dam, S., *DoD Architecture Framework: A Guide to Applying System Engineering to Develop Integrated Executable Architectures*, BookSurge Publishing, 2006.

[Date03] Date, C. J., *An Introduction to Database Systems*, 8th Edition, Addison Wesley, 2003.

[Denn08] Dennis, A. et al., *Systems Analysis and Designs*, 4th Edition, Wiley, 2008.

[Dori95] Dori, D., "Object-Process Analysis: Maintaining the Balance between System Structure and Behavior," *Journal of Logic and Computation* 5(2), pp.227-249, 1995.

[Dori02] Dori, D., *Object-Process Methodology: A Holistic Systems Paradigm*, Springer Verlag, New York, 2002.

[Dori16] Dori, D., *Model-Based Systems Engineering with OPM and SysML*, Springer Verlag, New York, 2016.

[Elma10] Elmasri, R., *Fundamentals of Database Systems*, 6th Edition, Addison Wesley, 2010.

[Hoar85] Hoare, C. A. R., *Communicating Sequential Processes*, Prentice-Hall, 1985.

[Kend10] Kendall, K. et al., *Systems Analysis and Designs*, 8th Edition, Prentice Hall, 2010.

[Maie09] Maier, M. W., *The Art of Systems Architecting*, 3rd Edition, CRC Press, 2009.

[Miln89] Milner, R., *Communication and Concurrency*, Prentice-Hall, 1989.

[Miln99] Milner, R., *Communicating and Mobile Systems: the π-Calculus*, 1st Edition, Cambridge University Press, 1999.

[O'Dri15] O'Driscoll, G., *Essential Guide to Smart Home Automation Safety & Security: Use Home Automation to Increase Your Families Safety Levels*, CreateSpace Independent Publishing Platform, 2015.

[O'Rou03] O'Rourke, C. et al, *Enterprise Architecture Using the Zachman Framework*, 1st Edition, Course Technology, 2003.

[Pele00] Peleg, M. et al., "The Model Multiplicity Problem: Experimenting with Real-Time Specification Methods". *IEEE Tran. on Software Engineering*. 26 (8), pp. 742–759, 2000.

[Prat00] Pratt, T. W. et al., *Programming Languages: Designs and Implementation*,

122

4th Edition, Prentice Hall 2000.

[Pres09] Pressman, R. S., *Software Engineering: A Practitioner's Approach*, 7th Edition, McGraw-Hill, 2009.

[Putm00] Putman, J. R. et al., *Architecting with RM-ODP*, Prentice-Hall, 2000.

[Rayn09] Raynard, B., *TOGAF The Open Group Architecture Framework 100 Success Secrets*, Emereo Pty Ltd, 2009.

[Roza11] Rozanski, N. et al., *Software Systems Architecture: Working With Stakeholders Using Viewpoints and Perspectives*, 2nd Edition, Addison-Wesley Professional, 2011.

[Rumb91] Rumbaugh, J. et al., *Object-Oriented Modeling and Design*, Prentice-Hall, 1991.

[Seth96] Sethi, R., *Programming Languages: Concepts and Constructs*, 2nd Edition, Addison-Wesley, 1996.

[Sode03] Soderborg, N.R. et al., "OPM-based Definitions and Operational Templates," *Communications of the ACM* 46(10), pp. 67-72, 2003.

[Somm06] Sommerville, I., *Software Engineering*, 8th Edition, Addison-Wesley, 2006.

[Toga08] The Open Group, *TOGAF Version 9 - A Manual (TOGAF Series)*, 9th Edition, Van Haren Publishing, 2008.

[Your99] Yourdon, E., *Death March: The Complete Software Developer's Guide to Surviving Mission Impossible Projects*, Prentice-Hall, 1999.